Complex Numbers
The Higher
Dimensional Forms

Second Edition

The Natural Algebras
The Nature of Space
The Algebraic Vector Product Spaces
The Abelian Group Algebras
The Unification of Groups, Algebra, and Geometry

Dennis Morris

First published in the UK in 2007 by
Abane & Right
31/32 Long Row
Port Mulgrave
Whitby
TS13 5LF

2nd Edition published in February 2015
Revised August 2015

Copyright © Dennis Morris

The moral right of Dennis Morris to be identified as the author of this work has been asserted by him in accordance with the Copyright, Designs and Patents Act 1988.

All rights reserved. No part of this publication may be reproduced or transmitted in any form or by any means, electronic or mechanical including photocopying, recording or any information storage or retrieval system, without prior permission in writing from the publisher.

Contents

Introduction to the 2nd Edition 1
Preliminaries 3
Historical Note 3
Note on the Hurwitz Theorem 5
Section 1 6
    Introduction to Section 1: 6
    The Matrix Representation of Algebras: 6
    The Nature of Algebraic Isomorphism: 9
    Overblown Representations: 12
    Algebraic Field Extensions: 13
    Upon Norms and Inner Products: 14
    Exponentiation of Matrices: 15
Section 2 16
    Introduction to Section 2: 16
    Complex Numbers by Matrices: 16
    Matrix Differentiation and the Cauchy Riemann Equations: .. 18
    Matrix Differentiation: 19
    Summary of differentiation: 23
    The Polar Form of Complex Numbers: 23
    Vectors and Inner Products: 27
    Summary: 30
    Logarithms and Powers: 32
    Vector calculus in 2-dimensional space: 35
Section 3 37
    Introduction to Section 3: 37
    The Hyperbolic Complex Numbers – Basics: 37
    The C-R Eqns. for Hyperbolic Complex Numbers: 41
    The Polar Form of Hyperbolic Complex Numbers: 41
    Logarithms and Powers: 43
    Odd Parity Matrices, Quaternions, and Dihedralions: 45
    Groups associated with these matrix forms: 54
    Hyperbolic Complex Numbers are not algebraically closed: . 55
    Vectors within Hyperbolic Space: 55
    Differentiation of fields - hyperbolic complex numbers: 56
Section 4 58
    Introduction to Section 4: 58
    1-dimensional Space: 58

The General Rotation Matrix – 2-dimensions: 60
Section 5 62
  Introduction to Section 5: 62
  The Algebraic Matrix Forms from the Field Axioms: 62
  The Axioms of an Algebraic Matrix Field: 62
  Block Algebras: 74
  Algebraic Matrix Forms by Cayley Tables: 77
  Illegitimate Spaces: 79
Section 6 84
  Introduction to Section 6: 84
  The Definition of the Trigonometric Functions: 84
  Definition of Simple Trigonometric Functions: 88
  Definition of Compound Trigonometric Functions: 92
  The Definition of Rotation: 93
  The Definition of the Distance Function: 96
  The Definition of Natural Space: 97
Section 7 99
  Introduction to Section 7: 99
  The Polar Form of the $C_3L^1H^2_{(j=1,k=1)}$ Algebra: 99
  The Polar Form of the $C_3L^1E^2_{(j=1,k=-1)}$ Algebra: 104
  Polar Forms $C_3L^1E^1H^1_{(j=-1,k=1)}$ and $C_3L^1E^1H^1_{(j=-1,k=-1)}$ Algebras: ... 108
  The 3-dimensional v-functions - an Introduction: 108
  The General Polar Form: 113
  The C-R Equations for the $C_3L^1H^2_{(j=1,k=1)}$ Algebra: 115
  Non-Matrix Notation of the Algebras: 117
  The 2-dimensional Shadows of 3-dimensional Algebras: 122
  Logarithms - $C_3L^1H^2_{(j=1,k=1)}$: 125
  3-dimensional Vectors: 126
Section 8 129
  Introduction to Section 8: 129
  The 3-dimensional Simple-Trig Functions: 129
  Differentiation: 131
  In Standard Function Form: 131
  Graphs: 132
  The $C_3L^1H^2_{(j=1,k=1)}$ v -Functions: 134
  Differentiation: 135
  3D-Graphs: 137

Section 9 138
Introduction to Section 9: 138
The Polar Form of the $C_{4,1}L^1H^3_{(j=-1,k=-1,l=-1)}$ Algebra: 138
The Polar Form of the $C_{4,1}L^1E^3_{(j=-1,k=-1,l=-1)}$ Algebra: 142
The Polar Forms of the $\{C_{4,1}, C_{4,2}, C_{4,3}\}$ Algebras in General: . 145
The Polar Forms of the $C_2 \times C_2$ Algebras: 146
The General Polar Form $C_{4,1}$: 149
Non-matrix Notation - $C_{4,1}L^1H^3_{(j=-1,k=-1,l=-1)}$: 149
The Euclidean shadow of the $C_{4,1}L^1H^3_{(j=-1,k=-1,l=-1)}$ Algebra: 152
The Cauchy Riemann Equations - $C_{4,2}L^1H^3_{(j=-1,k=-1,l=-1)}$: 153
Section 10 156
Introduction to Section 10: 156
The 4-dimensional Simple-trig Functions: 156
The ν-functions - $C_{4,1}L^1H^3_{(j=-1,k=-1,l=-1)}$: 159
The ν-functions - $C_{4,2}L^1H^3_{(j=-1,k=-1,l=-1)}$: 164
The ν-functions - $C_{4,1}L^1E^3_{(j=-1,k=-1,l=-1)}$: 164
Section 11 167
Introduction to Section 11: 167
Permutation Matrices: 167
Algebraic Isomorphism between Algebraic Matrix Forms: . 173
Permutation Matrices with Negative Entries: 176
Section 12 179
Introduction to Section 12: 179
The Folding Operation: 179
Unfolding Negative Real Numbers: 183
A Chain of Algebras: 184
Unfolding the Euclidean Complex Numbers: 186
Unfolding in General: 189
A Farewell to Negative Numbers: 191
Concluding Remarks 193
Appendix 1 194
Axioms of a Metric: 194
Axioms of an Inner Product: 194
Appendix 2 195
The Field Axioms: 195

- Within this Work:..196
- The Axioms of an Algebraic Matrix Field:..........................196
- Appendix 3 196
  - General Flat Algebraic Matrix Fields: 197
  - 2 Dimensions: 197
  - 3 Dimensions: 198
  - 4 Dimensions: 198
  - 5 Dimensions: 202
  - 6 Dimensions: 203
- Appendix 4 204
  - Circulant Matrices:..204
- Appendix 5.1 207
  - Polar Forms of the $C_{4.1}$ Algebras:......................................207
- Appendix 5.2 213
  - Polar Forms of the $C_2 \times C_2$ Algebras: 213
- Glossary & Nomenclature..219
- Notation..225
- Other Books by the Same Author 228
- Index...231

Introduction to the 2nd Edition

It is now eight years since the first edition of "Complex Numbers The Higher Dimensional Forms" was published. To your author's knowledge, the book has enjoyed sales on every continent except Antarctica. Even so, only some 500 copies have been sold to date.

The first edition of "Complex Numbers The Higher Dimensional Forms" was the first step in developing a whole new area of mathematics. The year 2015 has seen the publication by your author of two more books that develop this area of mathematics much further[1]. These books are:

"The Physics of Empty Space" by Dennis Morris

"The Naked Spinor (A Rewrite of Clifford Algebra)" by Dennis Morris

It seems that we have stumbled upon a revolutionary understanding of empty space, and interest in this area of mathematics is now starting to increase rapidly; there is thus a much increased demand for copies of "Complex Numbers The Higher Dimensional Forms". There are insufficient remaining printed copies of the first edition to meet the expected demand, and so it has been decided to issue a second edition. It has been decided to issue this 2nd edition in electronic book format for ease of international marketing, to avoid large up-front expenditure, and to cope with the vagaries of demand.

Unfortunately, the final proofs of the printed 1st edition of the book have been lost, and so this 2nd edition is based on proofs of an earlier draft. The reader will find small mismatches with the original 1st edition. We have also corrected a few typographical errors and tidied the manuscript a little.

This area of mathematics is now much better understood that it was when the 1st edition was printed. We now know that all forms of complex number, that is numbers with one real axis and $(n-1)$ imaginary axes, are spinor spaces and that the $\mathbb{R}^n$ types of space arise from these spinor spaces by superimposition of isomorphic algebras. We now know that the higher dimensional complex numbers that are of interest to the physicist are the non-commutative algebras that derive from the $C_2 \times C_2 \times ...$ finite groups. Unfortunately, such non-commutative algebras were not considered in the 1st

---

[1] Further books have been published in late 2015 and 2016 – See list at end of this book.

Introduction to 2nd Edition

edition of "Complex Numbers The Higher Dimensional Forms", or will be in this edition. This does not mean that this book "Complex Numbers The Higher Dimensional Forms" is not of interest anymore. This book fills out the foundations of the later work which concentrates upon the $C_2 \times C_2 \times ...$ groups. This book is much more mathematically thorough than the two later books noted above. One cannot claim to have a complete understanding of spinor theory and of empty space based on only the $C_2 \times C_2 \times ...$ groups, even though these do seem to be the physically important ones.

We have included many 2nd edition remarks and 2nd edition footnotes in this 2nd edition. As with any new area of mathematics, it gets pruned and shaped from its roughhewn form. So it has been with this area of mathematics. The 2nd edition remarks show this pruning and reshaping. Some of the material in the 1st edition has withered and is not used today, but one is loath to discard it completely for we never know when we might need this material. Some of the material in the 1st edition has expanded in unexpected directions. Some of the nomenclature has changed.

The 1st edition is rapidly becoming a collector's item. There are still a few copies left, and these can be bought via Amazon if the reader so wishes.

Dennis Morris (2015)

Preliminaries

Historical Note

The association between complex numbers and positions in 2-dimensional Euclidean space begins with Wallis (1616–1703) in 1673. It was subsequently stated by Wessel (1745–1818) in 1798 and again by Gauss (1777–1855) in 1799. It was given again by Argand (1768–1822) in 1806, and the association subsequently became known as the Argand diagram. In those times, Euclidean geometry was very prominent within mathematics and complex numbers were seen as being associated with the Euclidean geometry of 2-dimensions. It was inevitable that mathematicians would seek some higher kind of complex numbers associated with 3-dimensional Euclidean space, and it was expected that such numbers would exist.

Throughout all these times, mathematicians assumed, without question, that 2-dimensional space is necessarily a sub-space of 3-dimensional space. From this assumption, it follows that the 2-dimensional complex numbers are a sub-algebra of the anticipated 3-dimensional complex numbers and the way forward is to expand the complex numbers upwards into higher dimensions. These efforts were encouraged by some small, but tantalising, success and some almost success. In 1813, Servois[2] proposed a 3-dimensional kind of complex numbers that nearly worked, and in 1843 Hamilton (1805–1865) discovered the '4-dimensional' quaternions.

The quaternions work, they include the 2-dimensional complex numbers as a sub-algebra, but they are not an algebraic field because they are non-commutative. They are a division algebra. However, it was 'immediately obvious' to mathematicians of the day that higher dimensional kinds of complex numbers would be multiplicatively non-commutative because rotations within 3-dimensional Euclidean space are non-commutative. Tantalisingly, quaternions can be used, clumsily, to calculate rotations in 3-dimensional Euclidean space. By such tantalisation are human beings persuaded to believe the 'immediately obvious'.

Unfortunately, developing higher dimensional kinds of complex numbers was too hard, and so, in the middle of the 19th century, mathematicians developed a new approach to representing space. They turned their attention

---

[2] Your author is unaware of the years of Servois's birth and death.

to describing higher dimensional spaces by vectors. Vector algebra is such that 2-dimensional vector space is a sub-space of 3-dimensional vector space. Grassman (1809–1877) began this approach in 1844, but no one could read his work, and it was 1862, when he rewrote the work in a more readable form, before his work was appreciated. The vector algebra was expanded and deepened subsequently by Gibbs[3] (1839–1903) in 1881 and by Peano (1858–1932) in 1888.

Vector algebra underlies large areas of our understanding of the physical universe. It has been very successful, and is beloved by physicists and engineers. It is unfortunate that it is not an algebra. From a mathematical point of view, vector algebra is a mess. It is a linear space together with an undeclared metric together with two multiplication operations (the dot product and the cross product), neither of which are a bona-fide algebraic multiplication operation, and, if that's not bad enough, it has no vector product (algebraic multiplication operation).

Grassman was aware of this failing, and, from the start, he sought to introduce the exterior product of two vectors into the vector algebra. In 1878 and 1882 (posthumously) William K. Clifford (1845–1879) also attempted to introduce a vector product, known today as the Clifford product, into the vector algebra. In doing so, he invented the Clifford algebras. Unfortunately, neither Grassman's exterior product nor the Clifford product is an algebraic multiplication operation; it is not closed; neither the exterior product of two vectors nor the Clifford product of two vectors is a vector[4].

---

[3] It was Gibbs who took the Clifford product of quaternions and separated it into the dot product and cross product with which we are today familiar.
[4] But see "The Naked Spinor" by Dennis Morris

Note on the Hurwitz Theorem

The Hurwitz theorem is often stated as:

Hurwitz Theorem:
The only normed algebras over the reals are isomorphic to $\mathbb{R}, \mathbb{C}, \mathbb{H}, \mathbb{K}$.

This is slightly misleading in that this theorem covers only norms that have a Euclidean (with all positive signature) quadratic form similar to: $\sqrt{a^2 + b^2 + c^2}$. It is the assumption that 2-dimensional space is a sub-space of all the higher dimensional spaces that justifies covering only norms of this form. In this work, we will find many algebras over the reals that are normed, but the norm will not be a Euclidean quadratic form.

Section 1

Introduction to Section 1:

In this section, we write of mathematical concepts. The concepts will often be familiar, but we view them from a different angle. These changes are not changes in any logically deduced mathematics. We are not undoing mathematical proofs. These are changes in viewpoint. They are changes in the way we see numbers, algebras, and inner products. It is a change of understanding that is forced upon us by the mathematics that follows in later sections of the work. The reader might initially feel the changes in viewpoint are unnecessary or unjustified, and we make little attempt to justify them in this section. The necessity of these changes is to be found in the later sections of this work.

The Matrix Representation of Algebras:

An algebra is a mathematical construct that satisfies thirteen of the fourteen algebraic field axioms; the exception is multiplicative commutativity – see Appendix 2. If all fourteen axioms are satisfied, we have an algebraic field. It is with algebraic fields that we are concerned in this book[5].

Algebraic field is just a posh name for a type of numbers. The real numbers are an algebraic field as are the complex numbers.

An algebra is a linear space[6] together with an appropriate multiplication operation. A linear space is a set of linear transformations. Every linear transformation can be represented as a matrix. So every linear space is a particular set of matrices. We often have the view that a linear space is a set of mathematical objects like real numbers or complex numbers or magic squares. It is. However, a real number or a complex number or a magic square is a linear transformation.

The multiplication operation that we impose upon a linear space to produce an algebra does not have to be matrix multiplication, but, if the imposed multiplication operation is matrix multiplication, then we can do the whole of that algebra in matrix notation. Matrix multiplication seems to be the

---

[5] 2nd edition note: The two later books mentioned in the introduction to the second edition are concerned with non-commutative algebras. It is these non-commutative algebras that are of primary interest to the physicist.

[6] Linear spaces are also known as vector spaces – two names, same thing.

natural multiplication operation to impose upon a set of matrices. We shall therefore refer to linear spaces that have matrix multiplication for the multiplication operation as the natural algebras.[7]

Nomenclature:

Natural Algebra

An algebra that has matrix multiplication as the multiplication operation.

Natural Algebraic Field

An algebraic field that has matrix multiplication as the multiplication operation.

The form of multiplication that we call matrix multiplication is derived from the manipulation of linear equations. Thus, matrix multiplication operation could be equally well named linear multiplication. Using such parlance, we would define a natural algebra as a linear space together with linear multiplication. We note that linear addition is within the linear space definition.

Many multiplication operations are matrix multiplication in disguise. For example, multiplication of real numbers is matrix multiplication of 1×1 matrices. The multiplication operation of complex numbers is also matrix multiplication. We jump the gun to demonstrate[8].

The 2×2 matrix form of a complex number, $(a + \hat{i}b)$, is:

$$\begin{bmatrix} a & b \\ -b & a \end{bmatrix} \tag{1.1}$$

where the elements are all real numbers.

We have:

$$(a + \hat{i}b)(c + \hat{i}d) = (ac - bd) + (ad + bc)\hat{i} \tag{1.2}$$

Which is the same as:

$$\begin{bmatrix} a & b \\ -b & a \end{bmatrix} \begin{bmatrix} c & d \\ -d & c \end{bmatrix} = \begin{bmatrix} ac - bd & ad + bc \\ -(ad + bc) & ac - bd \end{bmatrix}$$

$$\begin{bmatrix} 0 & 1 \\ -1 & 0 \end{bmatrix}^2 = \begin{bmatrix} -1 & 0 \\ 0 & -1 \end{bmatrix} \tag{1.3}$$

---

[7] 2nd edition note: The reader should note that the term "Natural algebras" has not become accepted terminology and is not used today.

[8] Throughout this work, $\hat{i} = \sqrt[2]{-1}$ will have little hat on, as will similar objects.

Section 1

This is not the only matrix form of the complex numbers. The 4×4 matrix form with real numbers as elements:

$$\begin{bmatrix} a & 0 & c & 0 \\ 0 & a & 0 & c \\ -c & 0 & a & 0 \\ 0 & -c & 0 & a \end{bmatrix} \tag{1.4}$$

is also a representation of the complex numbers[9]:

$$\begin{bmatrix} a & 0 & c & 0 \\ 0 & a & 0 & c \\ -c & 0 & a & 0 \\ 0 & -c & 0 & a \end{bmatrix} \begin{bmatrix} e & 0 & g & 0 \\ 0 & e & 0 & g \\ -g & 0 & e & 0 \\ 0 & -g & 0 & e \end{bmatrix}$$

$$= \begin{bmatrix} ae-cg & 0 & ag+ce & 0 \\ 0 & ae-cg & 0 & ag+ce \\ -(ag+ce) & 0 & ae-cg & 0 \\ 0 & -(ag+ce) & 0 & ae-cg \end{bmatrix} \tag{1.5}$$

And:

$$\begin{bmatrix} 0 & 0 & 1 & 0 \\ 0 & 0 & 0 & 1 \\ -1 & 0 & 0 & 0 \\ 0 & -1 & 0 & 0 \end{bmatrix}^2 = \begin{bmatrix} -1 & 0 & 0 & 0 \\ 0 & -1 & 0 & 0 \\ 0 & 0 & -1 & 0 \\ 0 & 0 & 0 & -1 \end{bmatrix} \tag{1.6}$$

We emphasize that part of the definition of the algebra is that the elements within the matrix are real numbers.

An Aside:
   In the case where matrix elements are complex numbers, the block multiplication property of matrices allows us to write the matrix as a larger matrix of real numbers. For example, consider the matrix:

---

[9] 2nd edition note: It is part of a quaternion, and so it is not a true representation of the complex numbers – it has double cover.

$$\begin{bmatrix} a+\hat{i}b & c+\hat{i}d \\ e+\hat{i}f & g+\hat{i}h \end{bmatrix} \quad (1.7)$$

The block multiplication property of matrices allows us to write the above 2×2 matrix as a 4×4 matrix with real numbers as elements:

$$\begin{bmatrix} \begin{bmatrix} a & b \\ -b & a \end{bmatrix} & \begin{bmatrix} c & d \\ -d & c \end{bmatrix} \\ \begin{bmatrix} e & f \\ -f & e \end{bmatrix} & \begin{bmatrix} g & h \\ -h & g \end{bmatrix} \end{bmatrix} = \begin{bmatrix} a & b & c & d \\ -b & a & -d & c \\ e & f & g & h \\ -f & e & -h & g \end{bmatrix} \quad (1.8)$$

There is a thing of weight here. We no longer need the concept of an imaginary number.

Throughout this work, we will be concerned only with algebras that have matrix multiplication as the multiplication operation – the natural algebras. We will be concerned only with every such algebra that is an algebraic field. Thus, we will be concerned with only every natural algebraic field.

The Nature of Algebraic Isomorphism:

Two algebras over the same algebraic field[10] are said to be algebraically isomorphic if the operations of addition, multiplication, and multiplication by a scalar have the same effect upon all elements in both algebras. To be more technical, two algebras over the same field are algebraically isomorphic if:

1) There is a bi-jective (one-to-one and onto) mapping from the elements of the first algebra to the elements of the second algebra.
2) The sum of any two elements within the first algebra is the element that corresponds, under the bi-jective mapping, to the sum of the corresponding two elements, under the bi-jective mapping, in the second algebra.
3) The product of any two elements within the first algebra is the element that corresponds, under the bi-jective mapping, to the product of the corresponding two elements, under the bi-jective mapping, in the second algebra.
4) The result of multiplying any element within the first algebra by a scalar is the element that corresponds, under the bi-jective mapping, to the result of multiplying the corresponding element, under the bi-jective mapping, within the second algebra by the same scalar.

---

[10] Within this work, the algebraic field is always the real numbers.

Section 1

It is important to realise that the three operations mentioned above are the only operations that are bound to give the same results in both algebras. For example, the square-rooting operation need not give the same result in both algebras.

Definition:

Algebraic Isomorphism

Two algebras $\{A_1, A_2\}$ over the same field, F, are isomorphic if there is a bijective mapping from the first algebra to the second algebra such that:
$$\phi : \phi(s) \to s' \quad \forall s, t \in A_1 \quad \forall s't' \in A_2$$
i) $\phi(s+t) = s' + t'$
ii) $\phi(st) = s't'$
iii) $\phi(\alpha t) = \alpha s' \quad : \alpha \in F$

Theorem:
The algebra of matrices of the form:
$$\{a_{11} \in \mathbb{R}, a_{12} = 0, a_{21} = 0, a_{22} = a_{11}\} \equiv \begin{bmatrix} a & 0 \\ 0 & a \end{bmatrix} \quad (1.9)$$
is algebraically isomorphic to the algebra of the real numbers.

Proof:

$$\phi : (\mathbb{R}, +, \times) \mapsto (M_{(2,\mathbb{R})}, +, \bullet) \quad : \quad \phi(a) \mapsto \begin{bmatrix} a & 0 \\ 0 & a \end{bmatrix}$$

$$\phi(a+e) \mapsto \begin{bmatrix} a+e & 0 \\ 0 & a+e \end{bmatrix} = \begin{bmatrix} a & 0 \\ 0 & a \end{bmatrix} + \begin{bmatrix} e & 0 \\ 0 & e \end{bmatrix}$$

$$\phi(a \times e) \mapsto \begin{bmatrix} ae & 0 \\ 0 & ae \end{bmatrix} = \begin{bmatrix} a & 0 \\ 0 & a \end{bmatrix} \bullet \begin{bmatrix} e & 0 \\ 0 & e \end{bmatrix}$$

$$\phi(\alpha a) \mapsto \begin{bmatrix} \alpha a & 0 \\ 0 & \alpha a \end{bmatrix} = \alpha \begin{bmatrix} a & 0 \\ 0 & a \end{bmatrix}$$

Although we do not give them, proofs of a similar nature apply to all (real element) matrices with leading diagonal elements equal and all other elements zero. Hence forward, we will refer to such matrices as real-number matrices.

Complex Numbers The Higher Dimensional Forms – 2nd Edition

> Nomenclature:
> *Real Number Matrices*
> Square matrices with real elements with all leading diagonal elements equal and all other elements zero[11].

> **Theorem:**
> The algebra of matrices of the form:
> $$\{a_{11} \in \mathbb{R}, a_{12} \in \mathbb{R}, a_{21} = -a_{12}, a_{22} = a_{11}\} \equiv \begin{bmatrix} a & b \\ -b & a \end{bmatrix} \quad (1.10)$$
> is isomorphic to the algebra of the complex numbers.

Proof:

$$\phi:(\mathbb{C},+,\times) \mapsto (M_{(2,\mathbb{R})},+,\bullet) \quad : \quad \phi(a+\hat{i}b) \mapsto \begin{bmatrix} a & b \\ -b & a \end{bmatrix}$$

$$\phi((a+\hat{i}b)+(e+\hat{i}f)) \mapsto \begin{bmatrix} a+e & b+f \\ -(b+f) & a+e \end{bmatrix} = \begin{bmatrix} a & b \\ -b & a \end{bmatrix} + \begin{bmatrix} e & f \\ -f & e \end{bmatrix}$$

$$\phi((a+\hat{i}b)\times(e+\hat{i}f)) \mapsto \begin{bmatrix} ae-bf & af+be \\ -(af+be) & ae-bf \end{bmatrix} = \begin{bmatrix} a & b \\ -b & a \end{bmatrix} \bullet \begin{bmatrix} e & f \\ -f & e \end{bmatrix}$$

$$\phi(\alpha(a+\hat{i}b)) \mapsto \begin{bmatrix} \alpha a & -\alpha b \\ \alpha b & \alpha a \end{bmatrix} = \begin{bmatrix} \alpha & 0 \\ 0 & \alpha \end{bmatrix} \begin{bmatrix} a & -b \\ b & a \end{bmatrix}$$

For the reader's convenience, we point out that:

$$1 \cong \begin{bmatrix} 1 & 0 \\ -0 & 1 \end{bmatrix}, \quad \hat{i} \cong \begin{bmatrix} 0 & 1 \\ -1 & 0 \end{bmatrix}$$

$$a+\hat{i}b \cong \begin{bmatrix} a & b \\ -b & a \end{bmatrix}, \quad a-\hat{i}b \cong \begin{bmatrix} a & -b \\ b & a \end{bmatrix} \quad (1.11)$$

Multiplying 2×2 matrices, or larger matrices, by a real number, which is a 1×1 matrix, is an operation that is generally accepted within mathematics. It is technically incorrect. Having now proved the (conventional) algebraic isomorphism of the real numbers and the $n\times n$ real number matrices, we will henceforward consider scalars to be real number matrices of the appropriate size. Since algebraic isomorphism includes the multiplication operation, we gain nothing by also including the operation of scalar multiplication. Thus,

---

[11] These matrices are also known as homothetic matrices.

in our view, within this work, there is no requirement of isomorphic scalar multiplication.

It is important to realize that only the operations of addition and multiplication are bound to give the same results in both algebras.

Overblown Representations:

In 1×1 matrix form, the positive real numbers have two square roots. In 2×2 matrix form, we have:

$$\begin{bmatrix} w & x \\ y & z \end{bmatrix} \begin{bmatrix} w & x \\ y & z \end{bmatrix} = \begin{bmatrix} w^2 + xy & x(w+z) \\ y(w+z) & xy+z^2 \end{bmatrix} = \begin{bmatrix} a & 0 \\ 0 & a \end{bmatrix} \quad (1.12)$$

In the 2×2 matrix form, the real numbers have as many square roots as there are solutions of the equations:

$$w^2 + xy = a$$
$$x(w+z) = 0$$
$$y(w+z) = 0 \quad (1.13)$$
$$xy + z^2 = a$$

There are an infinite number of such solutions, and hence an infinite number of square roots.

Another example of two matrix representations being algebraically isomorphic but of different size is the complex numbers.

$\begin{bmatrix} a & b \\ -b & a \end{bmatrix}$ is algebraically isomorphic to $\begin{bmatrix} a & 0 & b & 0 \\ 0 & a & 0 & b \\ -b & 0 & a & 0 \\ 0 & -b & 0 & a \end{bmatrix}$.

The 2×2 matrix form has two square roots with real numbers as elements, both of which are complex numbers. The 4×4 matrix form has more than this, some of which are not complex numbers. In both cases, we have algebraically isomorphic algebras that do not have the same number of square roots.

Which of these algebras is the 'true' form of the real numbers and the 'true' form of the complex numbers? We take the view that the true form is the one with no zeros in the matrix – the 1×1 matrices in the case of the real numbers and the 2×2 matrices in the case of the complex numbers. We refer to the larger matrices, with the unnecessary zeros, as overblown representations of the algebras.

> Nomenclature: *Overblown Representation*
> A matrix representation of an algebra by a matrix that is larger than is necessary to include the algebraic operations[12] is called an overblown representation of that algebra.

When we seek to discover the roots of real numbers within an algebra, as we did above, because of the presence of the extra zeros, overblown representations allow more solutions than the 'true' representation of the algebra does. We avert from these overblown representations because the 'true' representations contain everything that is of the 'algebraic essence' of the algebra.

Algebraic Field Extensions:

It is conventional wisdom that the complex numbers are an algebraic field extension of the real numbers associated with the monic minimum polynomial $x^2 + 1 = 0$. We will shortly meet the hyperbolic complex numbers. These are an algebraic field and thus of the same algebraic status as the complex numbers. They contain within them the 2×2 real number matrices as a sub-algebra, as, of course, do the complex numbers. Conventional wisdom has it that, even though they contain the real numbers within them, the hyperbolic complex numbers are not an algebraic field extension of the real numbers because they are not associated with a monic minimum polynomial that will not split into linear factors over the real numbers.

We reject this conventional wisdom. Indeed, we reject the concept of algebraic extension. Instead, we adopt the concept of algebraic inclusion. We take the view that the complex number algebra contains within it a sub-algebra that is algebraically isomorphic (if overblown) to the real numbers. Similarly, the hyperbolic complex numbers contain within them a sub-algebra that is algebraically isomorphic to the real numbers. We take the view that un-splittable polynomials are nothing to do with algebraic inclusion[13].

---

[12] The algebraic operations are the operations of addition, multiplication, (and, conventionally, multiplication by a scalar).

[13] 2nd edition note: This change of view seems to upset a lot of people. Your author did not expect such upset. All we are doing is rejecting 300 years' worth of established mathematics.

Section 1

We will soon meet the 3-dimension complex numbers (the 3-dimensional natural algebras[14]). Like the lesser dimensional natural algebras, these algebras have within them a sub-algebra that is algebraically isomorphic to the real numbers. They do not include a sub-algebra that is algebraically isomorphic to any of the 2-dimensional natural algebras. We will soon meet the 4-dimension complex numbers. Like the lesser dimensional natural algebras, these algebras have within them a sub-algebra that is algebraically isomorphic to the real numbers. They also have a sub-algebra that is algebraically isomorphic to one of the 2-dimensional algebras.

This concept of algebraic inclusion rather than algebraic extension is a fundamental change to the way we view numbers. The conventional view is to begin with the natural integers and build upwards upon them. The existence of the hyperbolic complex numbers effectively topples that building. One cannot get to the hyperbolic complex numbers by building upwards from the real numbers through un-splittable polynomials, and yet they exist. The view now pressed upon us is that we have 1-dimensional numbers, the real numbers; we have 2-dimensional numbers; we have 3-dimensional numbers, and so on.

Upon Norms and Inner Products:

Most algebras are created by mathematicians. It works something like this:
1) Get up in the morning and invent some objects and an addition operation that form a linear space. We now have a linear space.
2) After breakfast, invent a multiplication operation to impose upon these objects. It can't be any old operation; it must be closed and have one or two other properties, but, other than this, the only limit upon it is the limit of the mathematician's imagination. We now have an algebra.
3) After lunch, invent a norm to put upon this algebra. Again, it can't be any old thing, but, within a few constraints, the only limit upon it is the limit of the mathematician's imagination. We now have a normed algebra.
4) After tea, invent an inner product operation to put upon this algebra. Again, within a few constraints, the only limit upon it is the limit of the mathematician's imagination. We now have a Hilbert space, and that is some baby of an algebra, and all invented by the mathematician, and all before dinner – no wonder the world so admires mathematicians.

Well, perhaps we've been a little flippant[15]. Creating an algebra is not quite that easy. The point is that the algebra, and all its bits, was invented and not

---

[14] A complex number algebra is the same thing as a natural algebra. We use the terms interchangeably.
[15] There's no 'perhaps' about it.

discovered. The natural algebras are different. Starting with the concept of linearity, writing the objects within the linear space as linear transformations (matrices), adopting the linear multiplication operation (matrix multiplication) creates the natural algebra. The natural algebra then takes control, as if it did not have control from the start. The natural algebra already has its norm within it, and the algebra imposes that norm upon the mathematician. Not only does it impose a norm upon the mathematician, but, with the norm, it also imposes a rotation matrix and a set of trigonometric functions. The natural algebra also has its inner product within it, and the algebra imposes that inner product upon the mathematician. By the same operation, the natural algebra forces (one or more) exterior product(s) on to the mathematician. We have turned algebra creation upon its head - the inmates have taken over the asylum.

Exponentiation of Matrices:

The material presented here is all standard stuff. It is presented to ensure that all readers are familiar with it. The exponential of a matrix is:

$$\exp([A]) = \frac{1}{0!} + \frac{[A]}{1!} + \frac{[A]^2}{2!} + \frac{[A]^3}{3!} + \ldots \quad (1.14)$$

However, we have:
$$\exp([A]+[B]) = \exp([A])\exp([B]) \quad (1.15)$$

only if:
$$[A][B] = [B][A] \quad (1.16)$$

And, importantly:
$$\det(\exp([A])) = \exp(\mathrm{Trace}[A]) \quad (1.17)$$

So:
$$\det\left(\exp\left(\begin{bmatrix} 0 & \sim \\ \sim & 0 \end{bmatrix}\right)\right) = 1 \quad (1.18)$$

Section 2

Introduction to Section 2:

In this section, we seek to familiarise the reader with the complex number algebra being done in matrix notation. We use the Euclidean complex numbers to unveil an example of a central theme of this work. That central theme is that a type of space is contained within each the natural algebras[16] – the unification of algebra and space.

Complex Numbers by Matrices:

The matrix representation of a complex number, $\left(a+\hat{i}b\right)$[17], is:

$$\begin{bmatrix} a & b \\ -b & a \end{bmatrix} \quad (2.1)$$

where every element is a real number[18]. It is important to realize that:

$$\begin{bmatrix} a & -b \\ b & a \end{bmatrix} \quad \begin{bmatrix} -a & -b \\ b & -a \end{bmatrix} \quad \begin{bmatrix} -a & b \\ -b & -a \end{bmatrix} \quad (2.2)$$

are matrices of the same form and are also complex numbers.

Nomenclature:
Algebraic Matrix Form
If a particular form of matrix is such that matrices of that form, together with matrix multiplication, are an algebra, then we will refer to that form of matrix as an 'algebraic matrix form'.

Addition is, of course, matrix addition:

$$\begin{bmatrix} a & b \\ -b & a \end{bmatrix} + \begin{bmatrix} c & d \\ -d & c \end{bmatrix} = \begin{bmatrix} a+c & b+d \\ -(b+d) & a+c \end{bmatrix} \quad (2.3)$$

The reader should note that the form of the matrix is maintained by addition. Multiplication is matrix multiplication:

$$\begin{bmatrix} a & b \\ -b & a \end{bmatrix} \begin{bmatrix} c & d \\ -d & c \end{bmatrix} = \begin{bmatrix} ac-bd & ad+bc \\ -(ad+bc) & ac-bd \end{bmatrix} \quad (2.4)$$

---

[16] 2nd edition note: We now take the view that the algebra and the space are the same thing.
[17] We put a little hat on our *i* to be consistent with notation used later.
[18] Proof of algebraic isomorphism is given in a previous chapter.

The reader should note that the form of the matrix is maintained by multiplication. The reader might like to compare the elements of the product matrix to complex number multiplication in non-matrix notation:

$$\left(a+\hat{i}b\right)\left(c+\hat{i}d\right)=\left(ac-bd\right)+\hat{i}\left(ad+bc\right) \tag{2.5}$$

Because the complex numbers are 2×2 matrices, they each have a determinant:

$$\det\left(\begin{bmatrix} a & b \\ -b & a \end{bmatrix}\right) = a^2 + b^2 \tag{2.6}$$

The reader should note that the determinant of a complex number is the same as the modulus (*norm*) of the complex number $\left(a+\hat{i}b\right)$.

The inverse of a complex number is:

$$\begin{bmatrix} a & b \\ -b & a \end{bmatrix}^{-1} = \frac{1}{a^2+b^2}\begin{bmatrix} a & -b \\ b & a \end{bmatrix} \tag{2.7}$$

The conjugate of a complex number in matrix form is the adjoint matrix of that complex number[19]. The adjoint of a matrix is the inverse matrix multiplied by the determinant of that matrix:

$$\begin{aligned}
Conjugate\left(\begin{bmatrix} a & b \\ -b & a \end{bmatrix}\right) &= \left(a^2+b^2\right)\begin{bmatrix} a & b \\ -b & a \end{bmatrix}^{-1} \\
&= \frac{a^2+b^2}{a^2+b^2}\begin{bmatrix} a & -b \\ b & a \end{bmatrix} \\
&= \begin{bmatrix} a & -b \\ b & a \end{bmatrix}
\end{aligned} \tag{2.8}$$

A Word of Warning:
Some texts say that the conjugate of a Euclidean complex number written in matrix notation is the transpose of that matrix. It is not so; the conjugate is the adjoint of that matrix, and it is accidental that, in the case of Euclidean complex numbers, the adjoint and the transpose are the same. This distinction is of no matter for Euclidean complex numbers but of great matter for the other algebras that we shall meet.

Division is by use of the conjugate, of course:

---

[19] 2nd edition note: We now take conjugation to be reverse rotation of the polar form of the algebra, which is what it always was anyway.

Section 2

$$\frac{\begin{bmatrix} a & b \\ -b & a \end{bmatrix}}{\begin{bmatrix} c & d \\ -d & c \end{bmatrix}} = \frac{\begin{bmatrix} a & b \\ -b & a \end{bmatrix}\begin{bmatrix} c & -d \\ d & c \end{bmatrix}}{\begin{bmatrix} c & d \\ -d & c \end{bmatrix}\begin{bmatrix} c & -d \\ d & c \end{bmatrix}}$$

$$= \frac{\begin{bmatrix} a & b \\ -b & a \end{bmatrix}\begin{bmatrix} c & -d \\ d & c \end{bmatrix}}{\begin{bmatrix} c^2+d^2 & 0 \\ 0 & c^2+d^2 \end{bmatrix}} \quad (2.9)$$

$$= \begin{bmatrix} \frac{1}{c^2+d^2} & 0 \\ 0 & \frac{1}{c^2+d^2} \end{bmatrix} \begin{bmatrix} a & b \\ -b & a \end{bmatrix}\begin{bmatrix} c & -d \\ d & c \end{bmatrix}$$

which one can compare to:

$$\frac{a+\hat{i}b}{c+\hat{i}d} = \frac{(a+\hat{i}b)(c-\hat{i}d)}{(c+\hat{i}d)(c-\hat{i}d)} = \frac{(a+\hat{i}b)(c-\hat{i}d)}{c^2+d^2} \quad (2.10)$$

We have:

$$\begin{bmatrix} a & b \\ -b & a \end{bmatrix} = \begin{bmatrix} a & 0 \\ 0 & a \end{bmatrix} + \begin{bmatrix} 0 & b \\ -b & 0 \end{bmatrix}$$

$$= \begin{bmatrix} a & 0 \\ 0 & a \end{bmatrix}\begin{bmatrix} 1 & 0 \\ 0 & 1 \end{bmatrix} + \begin{bmatrix} b & 0 \\ 0 & b \end{bmatrix}\begin{bmatrix} 0 & 1 \\ -1 & 0 \end{bmatrix} \quad (2.11)$$

and the reader might like to think of the diagonal matrix (the one with the *a*'s in it) as the real part of the complex number and the other matrix (the one with the *b*'s in it) as the imaginary part. The reader should note that we have no need of the imaginary number $\hat{i} = \sqrt{-1}$ (all the elements of the matrix are real numbers), but that we do have need of the matrix:

$$\begin{bmatrix} 0 & 1 \\ -1 & 0 \end{bmatrix}^2 = \begin{bmatrix} -1 & 0 \\ 0 & -1 \end{bmatrix} \quad (2.12)$$

Matrix Differentiation and the Cauchy Riemann Equations:

A function of a 2×2 matrix is a mapping from a 2×2 matrix to a 2×2 matrix of the form:

$$f\left(\begin{bmatrix} a & b \\ c & d \end{bmatrix}\right) \to \begin{bmatrix} u(a,b,c,d) & v(a,b,c,d) \\ s(a,b,c,d) & t(a,b,c,d) \end{bmatrix} \quad (2.13)$$

Complex Numbers The Higher Dimensional Forms – 2nd Edition

In this chapter, we consider only matrices of the form:

$$f\left(\begin{bmatrix} a & b \\ -b & a \end{bmatrix}\right) \to \begin{bmatrix} u(a,b) & v(a,b) \\ -v(a,b) & u(a,b) \end{bmatrix} \qquad (2.14)$$

Matrix Differentiation:

In general, we do not know how to differentiate matrices of variables, but we do know how to differentiate real number variables. Because we know how to differentiate real number variables, we know how to differentiate matrix variables of the form:

$$\begin{bmatrix} u(a,b) & 0 \\ 0 & u(a,b) \end{bmatrix} \qquad (2.15)$$

The reader will recall that matrices of this form are isomorphic to the real numbers. We approach the differentiation of matrices of the form:

$$\begin{bmatrix} u(a,b) & v(a,b) \\ -v(a,b) & u(a,b) \end{bmatrix} \qquad (2.16)$$

by splitting them and doing a bit of matrix arithmetic:

$$\begin{bmatrix} u(a,b) & v(a,b) \\ -v(a,b) & u(a,b) \end{bmatrix} = \begin{bmatrix} u(a,b) & 0 \\ 0 & u(a,b) \end{bmatrix} + \begin{bmatrix} 0 & v(a,b) \\ -v(a,b) & 0 \end{bmatrix}$$

$$= \begin{bmatrix} u(a,b) & 0 \\ 0 & u(a,b) \end{bmatrix} + \begin{bmatrix} 0 & 1 \\ -1 & 0 \end{bmatrix} \begin{bmatrix} v(a,b) & 0 \\ 0 & v(a,b) \end{bmatrix}$$

(2.17)

This is the sum of two real number variables where the matrix $\begin{bmatrix} 0 & 1 \\ -1 & 0 \end{bmatrix}$ is a constant.

Technique[20]:
Matrix Differentiation
You take a matrix. You split it into pieces. You turn each piece into a real number matrix and a constant. Then you can differentiate it. Clearly, this will work only with 'special' types of matrix.

We will be differentiating with respect to a real number variable. This means that we differentiate with respect to either:

---

[20] See page 35.

Section 2

$$\begin{bmatrix} a & 0 \\ 0 & a \end{bmatrix} \text{ or } \begin{bmatrix} 0 & 1 \\ -1 & 0 \end{bmatrix} \begin{bmatrix} b & 0 \\ 0 & b \end{bmatrix}$$

We consider functions of complex numbers. These are of the form:

$$f\left(\begin{bmatrix} a & b \\ -b & a \end{bmatrix}\right) \to \begin{bmatrix} u(a,b) & v(a,b) \\ -v(a,b) & u(a,b) \end{bmatrix} \qquad (2.18)$$

Differentiation of the function $f\left(\begin{bmatrix} a & b \\ -b & a \end{bmatrix}\right)$ with respect to $\begin{bmatrix} a & 0 \\ 0 & a \end{bmatrix}$ at point $\begin{bmatrix} \alpha_{11} & -\alpha_{21} \\ \alpha_{21} & \alpha_{11} \end{bmatrix}$ is (holding b constant):

$$f'\left(\begin{bmatrix} \alpha_{11} & \alpha_{12} \\ -\alpha_{21} & \alpha_{11} \end{bmatrix}\right) =$$

$$\lim_{\begin{bmatrix} a & a_{21} \\ -a_{21} & a \end{bmatrix} \to \begin{bmatrix} \alpha_{11} & a_{21} \\ -a_{21} & \alpha_{11} \end{bmatrix}} \frac{\begin{bmatrix} u(a,\alpha_{21}) & v(a,\alpha_{21}) \\ -v(a,\alpha_{21}) & u(a,\alpha_{21}) \end{bmatrix} - \begin{bmatrix} u(\alpha_{11},\alpha_{21}) & v(\alpha_{11},\alpha_{21}) \\ -v(\alpha_{11},\alpha_{21}) & u(\alpha_{11},\alpha_{21}) \end{bmatrix}}{\begin{bmatrix} a & \alpha_{21} \\ -\alpha_{21} & a \end{bmatrix} - \begin{bmatrix} \alpha_{11} & \alpha_{21} \\ -\alpha_{21} & \alpha_{11} \end{bmatrix}} \qquad (2.19)$$

Now:

$$f'\left(\begin{bmatrix} \alpha_{11} & \alpha_{12} \\ -\alpha_{21} & \alpha_{11} \end{bmatrix}\right) =$$

$$\lim_{\begin{bmatrix} a & a_{21} \\ -a_{21} & a \end{bmatrix} \to \begin{bmatrix} \alpha_{11} & a_{21} \\ -a_{21} & \alpha_{11} \end{bmatrix}} \frac{\begin{bmatrix} u(a,\alpha_{21}) & 0 \\ 0 & u(a,\alpha_{21}) \end{bmatrix} - \begin{bmatrix} u(\alpha_{11},\alpha_{21}) & 0 \\ 0 & u(\alpha_{11},\alpha_{21}) \end{bmatrix}}{\begin{bmatrix} a & 0 \\ 0 & a \end{bmatrix} - \begin{bmatrix} \alpha_{11} & 0 \\ 0 & \alpha_{11} \end{bmatrix}}$$

$$+ \begin{bmatrix} 0 & 1 \\ -1 & 0 \end{bmatrix} \lim_{\begin{bmatrix} a & a_{21} \\ -a_{21} & a \end{bmatrix} \to \begin{bmatrix} \alpha_{11} & a_{21} \\ -a_{21} & \alpha_{11} \end{bmatrix}} \frac{\begin{bmatrix} v(a,\alpha_{21}) & 0 \\ 0 & v(a,\alpha_{21}) \end{bmatrix} - \begin{bmatrix} v(\alpha_{11},\alpha_{21}) & 0 \\ 0 & v(\alpha_{11},\alpha_{21}) \end{bmatrix}}{\begin{bmatrix} a & 0 \\ 0 & a \end{bmatrix} - \begin{bmatrix} \alpha_{11} & 0 \\ 0 & \alpha_{11} \end{bmatrix}}$$

$$(2.20)$$

And:

$$f'\left(\begin{bmatrix} \alpha_{11} & \alpha_{12} \\ -\alpha_{21} & \alpha_{11} \end{bmatrix}\right) = \begin{bmatrix} \dfrac{\partial u}{\partial a} & 0 \\ 0 & \dfrac{\partial u}{\partial a} \end{bmatrix} + \begin{bmatrix} 0 & 1 \\ -1 & 0 \end{bmatrix}\begin{bmatrix} \dfrac{\partial v}{\partial a} & 0 \\ 0 & \dfrac{\partial v}{\partial a} \end{bmatrix} \quad (2.21)$$

Which in non-matrix notation would be written as:

$$f'(\alpha_{11},\alpha_{21}) = \frac{\partial u}{\partial a}(\alpha_{11},\alpha_{21}) + \hat{i}\frac{\partial v}{\partial a}(\alpha_{11},\alpha_{21}) \quad (2.22)$$

We write it as:

$$f'\left(\begin{bmatrix} \alpha_{11} & \alpha_{12} \\ -\alpha_{21} & \alpha_{11} \end{bmatrix}\right) = \begin{bmatrix} \dfrac{\partial u}{\partial a}(\alpha_{11},\alpha_{21}) & \dfrac{\partial v}{\partial a}(\alpha_{11},\alpha_{21}) \\ -\dfrac{\partial v}{\partial a}(\alpha_{11},\alpha_{21}) & \dfrac{\partial u}{\partial a}(\alpha_{11},\alpha_{21}) \end{bmatrix} \quad (2.23)$$

We now repeat the procedure holding a constant.

Differentiation of the function $f\left(\begin{bmatrix} a & b \\ -b & a \end{bmatrix}\right)$ with respect to:

$$\begin{bmatrix} 0 & 1 \\ -1 & 0 \end{bmatrix}\begin{bmatrix} b & 0 \\ 0 & b \end{bmatrix} = \begin{bmatrix} 0 & b \\ -b & 0 \end{bmatrix} \quad (2.24)$$

at point $\begin{bmatrix} \alpha_{11} & \alpha_{21} \\ -\alpha_{21} & \alpha_{11} \end{bmatrix}$ is (holding a constant):

$$f'\left(\begin{bmatrix} \alpha_{11} & \alpha_{12} \\ -\alpha_{21} & \alpha_{11} \end{bmatrix}\right) = \lim_{\begin{bmatrix} a & b \\ -b & a \end{bmatrix} \to \begin{bmatrix} \alpha_{11} & \alpha_{21} \\ -\alpha_{21} & \alpha_{11} \end{bmatrix}} \frac{\begin{bmatrix} u(\alpha_{11},b) & v(\alpha_{11},b) \\ -v(\alpha_{11},b) & u(\alpha_{11},b) \end{bmatrix} - \begin{bmatrix} u(\alpha_{11},\alpha_{21}) & v(\alpha_{11},\alpha_{21}) \\ -v(\alpha_{11},\alpha_{21}) & u(\alpha_{11},\alpha_{21}) \end{bmatrix}}{\begin{bmatrix} \alpha_{11} & b \\ -b & \alpha_{11} \end{bmatrix} - \begin{bmatrix} \alpha_{11} & \alpha_{21} \\ -\alpha_{21} & \alpha_{11} \end{bmatrix}} \quad (2.25)$$

Now:

$$f'\left(\begin{bmatrix} \alpha_{11} & \alpha_{12} \\ -\alpha_{21} & \alpha_{11} \end{bmatrix}\right) = \quad (2.26)$$

21

Section 2

$$\lim_{\begin{bmatrix}a_{11} & b \\ -b & a_{11}\end{bmatrix} \to \begin{bmatrix}a_{11} & a_{21} \\ -a_{21} & a_{11}\end{bmatrix}} \frac{\begin{bmatrix}u(a_{11},b) & 0 \\ 0 & u(a_{11},b)\end{bmatrix} - \begin{bmatrix}u(a_{11},a_{21}) & 0 \\ 0 & u(a_{11},a_{21})\end{bmatrix}}{\begin{bmatrix}0 & b \\ -b & 0\end{bmatrix} - \begin{bmatrix}0 & a_{21} \\ -a_{21} & 0\end{bmatrix}}$$

$$+ \lim_{\begin{bmatrix}a_{11} & b \\ -b & a_{11}\end{bmatrix} \to \begin{bmatrix}a_{11} & a_{21} \\ -a_{21} & a_{11}\end{bmatrix}} \frac{\begin{bmatrix}0 & v(a_{11},b) \\ -v(a_{11},b) & 0\end{bmatrix} - \begin{bmatrix}0 & v(a_{11},a_{21}) \\ -v(a_{11},a_{21}) & 0\end{bmatrix}}{\begin{bmatrix}0 & b \\ -b & 0\end{bmatrix} - \begin{bmatrix}0 & a_{21} \\ -a_{21} & 0\end{bmatrix}}$$

(2.27)

We now change the variable that we are differentiating with respect to from $\begin{bmatrix}0 & b \\ -b & 0\end{bmatrix}$ to $\begin{bmatrix}0 & 1 \\ -1 & 0\end{bmatrix}\begin{bmatrix}b & 0 \\ 0 & b\end{bmatrix}$

$$f'\left(\begin{bmatrix}\alpha_{11} & \alpha_{12} \\ -\alpha_{21} & \alpha_{11}\end{bmatrix}\right) =$$

$$\lim_{\begin{bmatrix}a_{11} & b \\ -b & a_{11}\end{bmatrix} \to \begin{bmatrix}a_{11} & a_{21} \\ -a_{21} & a_{11}\end{bmatrix}} \frac{\begin{bmatrix}u(a_{11},b) & 0 \\ 0 & u(a_{11},b)\end{bmatrix} - \begin{bmatrix}u(a_{11},a_{21}) & 0 \\ 0 & u(a_{11},a_{21})\end{bmatrix}}{\begin{bmatrix}0 & 1 \\ -1 & 0\end{bmatrix}\left(\begin{bmatrix}b & 0 \\ 0 & b\end{bmatrix} - \begin{bmatrix}a_{21} & 0 \\ 0 & a_{21}\end{bmatrix}\right)}$$

(2.28)

$$+ \begin{bmatrix}0 & 1 \\ -1 & 0\end{bmatrix} \lim_{\begin{bmatrix}a_{11} & b \\ -b & a_{11}\end{bmatrix} \to \begin{bmatrix}a_{11} & a_{21} \\ -a_{21} & a_{11}\end{bmatrix}} \frac{\begin{bmatrix}v(a_{11},b) & 0 \\ 0 & v(a_{11},b)\end{bmatrix} - \begin{bmatrix}v(a_{11},a_{21}) & 0 \\ 0 & v(a_{11},a_{21})\end{bmatrix}}{\begin{bmatrix}0 & 1 \\ -1 & 0\end{bmatrix}\left(\begin{bmatrix}b & 0 \\ 0 & b\end{bmatrix} - \begin{bmatrix}a_{21} & 0 \\ 0 & a_{21}\end{bmatrix}\right)}$$

And:

$$f'\left(\begin{bmatrix}\alpha_{11} & \alpha_{12} \\ -\alpha_{21} & \alpha_{11}\end{bmatrix}\right) = \frac{1}{\begin{bmatrix}0 & 1 \\ -1 & 0\end{bmatrix}} \begin{bmatrix}\frac{\partial u}{\partial b} & 0 \\ 0 & \frac{\partial u}{\partial b}\end{bmatrix} + \begin{bmatrix}\frac{\partial v}{\partial b} & 0 \\ 0 & \frac{\partial v}{\partial b}\end{bmatrix}$$

$$= \begin{bmatrix}0 & -1 \\ 1 & 0\end{bmatrix}\begin{bmatrix}\frac{\partial u}{\partial b} & 0 \\ 0 & \frac{\partial u}{\partial b}\end{bmatrix} + \begin{bmatrix}\frac{\partial v}{\partial b} & 0 \\ 0 & \frac{\partial v}{\partial b}\end{bmatrix}$$

(2.29)

Which in non-matrix notation would be written as:

$$f'(\alpha_{11},\alpha_{21}) = \frac{\partial v}{\partial b}(\alpha_{11},\alpha_{21}) - \hat{i}\frac{\partial u}{\partial b}(\alpha_{11},\alpha_{21}) \qquad (2.30)$$

We write it as:

$$f'\left(\begin{bmatrix} \alpha_{11} & \alpha_{12} \\ -\alpha_{21} & \alpha_{11} \end{bmatrix}\right) = \begin{bmatrix} \frac{\partial v}{\partial b}(\alpha_{11},\alpha_{21}) & -\frac{\partial u}{\partial b}(\alpha_{11},\alpha_{21}) \\ \frac{\partial u}{\partial b}(\alpha_{11},\alpha_{21}) & \frac{\partial v}{\partial b}(\alpha_{11},\alpha_{21}) \end{bmatrix} \qquad (2.31)$$

The complex matrix function is differentiable only if the two differentiations are equal, ie:

$$\frac{\partial u}{\partial a} = \frac{\partial v}{\partial b} \quad \& \quad \frac{\partial u}{\partial b} = -\frac{\partial v}{\partial a} \qquad (2.32)$$

These are the familiar Cauchy-Riemann equations of complex analysis. The Cauchy Riemann equations say that we will get the same answer regardless of which variable we differentiate with respect to. This means that the space (Argand diagram) that we are working in is isotropic. It is the (subjective) insistence upon the isotropy of space that leads to the Cauchy Riemann equations, and so we are going around in circles.

Summary of differentiation:

Factor out the imaginary parts of the matrices to leave variables as only real number matrices. Differentiate and then multiply the imaginary parts back into the expression[21].

The Polar Form of Complex Numbers:

We have:

$$\begin{bmatrix} a & 0 \\ 0 & a \end{bmatrix}\begin{bmatrix} 0 & b \\ -b & 0 \end{bmatrix} = \begin{bmatrix} 0 & b \\ -b & 0 \end{bmatrix}\begin{bmatrix} a & 0 \\ 0 & a \end{bmatrix} \qquad (2.33)$$

Because these sub-matrices are multiplicatively commutative, we can split them and exponentiate them:

---

[21] 2nd edition note: Because in this book all matrices are commutative, it matters not to which side we factor the imaginary matrices out. Differentiation of non-commutative matrices is dealt with in "The Physics of Empty Space" by Dennis Morris.

Section 2

$$\exp\left(\begin{bmatrix} a & b \\ -b & a \end{bmatrix}\right) = \exp\left(\begin{bmatrix} a & 0 \\ 0 & a \end{bmatrix} + \begin{bmatrix} 0 & b \\ -b & 0 \end{bmatrix}\right)$$

$$= \exp\left(\begin{bmatrix} a & 0 \\ 0 & a \end{bmatrix}\right) \exp\left(\begin{bmatrix} 0 & b \\ -b & 0 \end{bmatrix}\right)$$

$$= \begin{bmatrix} e^a & 0 \\ 0 & e^a \end{bmatrix} \begin{bmatrix} 1 - \dfrac{b^2}{2!} + \dfrac{b^4}{4!} - \ldots & b - \dfrac{b^3}{3!} + \dfrac{b^5}{5!} - \ldots \\ -b + \dfrac{b^3}{3!} - \dfrac{b^5}{5!} + \ldots & 1 - \dfrac{b^2}{2!} + \dfrac{b^4}{4!} - \ldots \end{bmatrix}$$

$$= \begin{bmatrix} r & 0 \\ 0 & r \end{bmatrix} \begin{bmatrix} \cos b & \sin b \\ -\sin b & \cos b \end{bmatrix}$$

(2.34)

The algebra of the real numbers is associated with the 1-dimensional space that is the real number line. Thus, the real number matrix is a length matrix. The matrix with the trigonometric functions in it is the rotation matrix of 2-dimensional Euclidean space. Because the b sub-matrix has trace equal to zero, we have:

$$\det\left(\exp\left(\begin{bmatrix} 0 & b \\ -b & 0 \end{bmatrix}\right)\right) = \exp(0) = 1 \qquad (2.35)$$

and this is the origin of the identity:

$$\cos^2 b + \sin^2 b = 1 \qquad (2.36)$$

Because the complex numbers are closed under exponentiation[22], we have:

$$\begin{bmatrix} x & y \\ -y & x \end{bmatrix} = \begin{bmatrix} r & 0 \\ 0 & r \end{bmatrix} \begin{bmatrix} \cos b & \sin b \\ -\sin b & \cos b \end{bmatrix}$$

(2.37)

$$\det\left(\begin{bmatrix} x & y \\ -y & x \end{bmatrix}\right) = \det\left(\begin{bmatrix} r & 0 \\ 0 & r \end{bmatrix} \begin{bmatrix} \cos b & \sin b \\ -\sin b & \cos b \end{bmatrix}\right)$$

$$x^2 + y^2 = r^2$$

Leading to:

$$r = \sqrt{x^2 + y^2} \qquad (2.38)$$

which is the metric of 2-dimensional Euclidean space.

---

[22] Exponentiation is no more than multiplication and addition.

The determinant of a 2×2 matrix is, by definition, the area of the parallelogram formed by the two vectors that are the columns (or rows) of the matrix. Thus, we ought not to be surprised that the square root of the determinant is a distance. Similar considerations apply to all $n \times n$ matrices except that we deal analogously with volumes and cube roots or hyper-volumes and 4th roots and so on. Within this work, we will find that, in general, the distance function of an n-dimensional space is the nth root of the determinant of the algebraic matrix form associated with that space. We have:

$$\begin{bmatrix} x & y \\ -y & x \end{bmatrix} = \begin{bmatrix} r & 0 \\ 0 & r \end{bmatrix} \begin{bmatrix} \cos b & \sin b \\ -\sin b & \cos b \end{bmatrix}$$

$$\begin{bmatrix} \dfrac{x}{r} & \dfrac{y}{r} \\ -\dfrac{y}{r} & \dfrac{x}{r} \end{bmatrix} = \begin{bmatrix} \cos b & \sin b \\ -\sin b & \cos b \end{bmatrix} \quad (2.39)$$

The determinant of the rotation matrix is unity, and so the determinant of the $\left(\dfrac{x}{r}, \dfrac{y}{r}\right)$ matrix is also unity. We have normalised the $\{x, y\}$ matrix. We see that the polar form of a normalised matrix is such that the trigonometric functions are equal to the normalised values upon the axes:

$$\cos b = \frac{x}{r}$$
$$\sin b = \frac{y}{r} \quad (2.40)$$

In other words, we have it that the trigonometric functions are projections from the normalised position (x, y) on to the axes of the space. We knew this anyway. The cosine function is the projection from the unit circle on to the horizontal axis, and the sine function is the projection from the unit circle on to the vertical axis. However, if, as we will, we meet unfamiliar functions in place of the sine and cosine, the same procedure will confirm them to be projections from the unit 'sphere'[23] on to the different axes.

There are three points to be made:
1) The trigonometric functions associated with 2-dimensional Euclidean space fall out of the algebra of complex numbers.

---

[23] We use the word sphere in its general multi-dimensional sense that includes circles, 4-dimensional spheres, etc.

2) The rotation matrix of 2-dimensional Euclidean space falls out of the algebra of complex numbers

3) The metric of 2-dimensional Euclidean space falls out of the algebra of complex numbers.

We take the view that this is sufficient to say that 2-dimensional Euclidean space is contained within the algebra of the complex numbers[24]. Worth a theorem, methinks[25].

> **Theorem:**
> 2-dimensional Euclidean space is contained within the algebra of the complex numbers.
> **Proof:**
> See Above.

With this theorem, we have effectively unified the geometry of 2-dimensional Euclidean space and the algebra of the complex numbers. If we can show that complex numbers exist even if the universe does not exist[26], then we can show that space has no choice other than to exist.[27]

We point out that the distance function (the metric) is symmetric in $\{x, y\}$ by which we mean that, if $\{x, y\}$ are swapped around, the output of the distance function is unaffected. This means that 2-dimensional Euclidean space is isotropic[28]. In particular, we point out that the x-axis is of the same nature as the y-axis; if one is a spatial axis, then so is the other; if one is a temporal axis, then so is the other. Thus, the space that has fallen out of the complex numbers is isotropic. This leads to the necessity of the Cauchy Riemann equations[29].

---

[24] This is a view and not a mathematical deduction. The view depends upon just what one requires to make a space. Many mathematicians are willing to accept the metric only.

[25] 2nd edition note: We see here that the author is trying to persuade himself that the complex algebra and empty space are the same thing. He had his doubts then, but it is rapidly becoming the accepted view that complex algebras are the same thing as spinor spaces.

[26] Which we shall not be attempting to do.

[27] In the case of the hyperbolic complex numbers that we are yet to meet, we get space-time within the algebra.

[28] It is also homogeneous because homogeneity is a property of the real numbers which sit upon the axes.

[29] 2nd edition note: The Cauchy-Riemann equations seem to be of no importance in any later developments.

Vectors and Inner Products:

To be a vector, a mathematical object has to be one of a set of similar objects that satisfy the axioms of a linear space. Indeed, linear spaces are often called vector spaces. An algebraic field is a linear space together with a suitable multiplication operation, and so it ought to be no surprise that algebraic fields can do everything that vectors can do and plus some.

Within physics, we often find the use of mathematical, arrow-like, objects called vectors (usually written in boldface type or with a line underneath them) to describe various physical situations. These mathematical, arrow-like, objects, together with two operations called the inner product (also known as dot product or scalar product) and the cross product are referred to as 'vector algebra'. This is a terminological inexactitude. These arrow-like objects with these two operations is not an algebra, and neither of the two operations is a bona-fide algebraic operation. From a mathematical point of view, so-called vector algebra is a mess. The only reason physicists like vector algebra is because it works perfectly[30]. To make vector algebra into a proper algebra we need an algebraic vector product. An algebraic vector product is an appropriate[31] way of multiplying two vectors together to produce another vector.

There is one-to-one correspondence between the 2-dimensional vectors and the complex number matrices:

$$\begin{bmatrix} a & b \\ -b & a \end{bmatrix} \leftrightarrow \begin{bmatrix} a \\ b \end{bmatrix} \qquad (2.41)$$

The complex number matrix, as well as being a complex number, is a position vector in 2-dimensional Euclidean space[32]. Within the complex number algebra is the product of two complex numbers, which is a complex number of course. This is our algebraic vector product. The algebraic way to do vector algebra is to do it using square matrices that form an algebraic field[33].

Within traditional vector algebra, we have the inner product and the cross product. There is such an operation within the complex numbers. Within the complex numbers, this operation is no more than a means of calculating the

---

[30] And, for this price, they sell their souls - bunch of tarts.
[31] It is appropriate if it satisfies the algebraic field axioms.
[32] 2nd edition note: 2-dimensional Euclidean space is the complex plane, not $\mathbb{R}^2$.
[33] 2nd edition note: At this point in the first edition, we had a footnote saying we cannot form a polar form with non-commutative matrices. This was wrong and was a stupid thing to say – see other books by Dennis Morris.

Section 2

angle subtended at the origin between two vectors. Take two complex numbers in both Cartesian and polar forms:

$$\left\{ \begin{bmatrix} x & y \\ -y & x \end{bmatrix}, \begin{bmatrix} t & z \\ -z & t \end{bmatrix} \right\}$$

$$\left\{ \begin{bmatrix} r & 0 \\ 0 & r \end{bmatrix} \begin{bmatrix} \cos\chi & \sin\chi \\ -\sin\chi & \cos\chi \end{bmatrix}, \begin{bmatrix} s & 0 \\ 0 & s \end{bmatrix} \begin{bmatrix} \cos(\chi+\phi) & \sin(\chi+\phi) \\ -\sin(\chi+\phi) & \cos(\chi+\phi) \end{bmatrix} \right\} \quad (2.42)$$

where ϕ is the angle between the two complex numbers subtended at the origin. Normalize the matrices, take the conjugate of either one of the pair, and multiply the resulting pair together leading to:

$$\begin{bmatrix} \frac{x}{\sqrt{x^2+y^2}} & \frac{-y}{\sqrt{x^2+y^2}} \\ \frac{y}{\sqrt{x^2+y^2}} & \frac{x}{\sqrt{x^2+y^2}} \end{bmatrix} \begin{bmatrix} \frac{t}{\sqrt{t^2+z^2}} & \frac{z}{\sqrt{t^2+z^2}} \\ \frac{-z}{\sqrt{t^2+z^2}} & \frac{t}{\sqrt{t^2+z^2}} \end{bmatrix} \quad (2.43)$$

$$= \begin{bmatrix} \frac{xt+yz}{\sqrt{x^2+y^2}\sqrt{t^2+z^2}} & \frac{xz-yt}{\sqrt{x^2+y^2}\sqrt{t^2+z^2}} \\ -\left(\frac{xz-yt}{\sqrt{x^2+y^2}\sqrt{t^2+z^2}} \right) & \frac{xt+yz}{\sqrt{x^2+y^2}\sqrt{t^2+z^2}} \end{bmatrix} \quad (2.44)$$

And:

$$\begin{bmatrix} \cos\chi & -\sin\chi \\ \sin\chi & \cos\chi \end{bmatrix} \begin{bmatrix} \cos(\chi+\phi) & \sin(\chi+\phi) \\ -\sin(\chi+\phi) & \cos(\chi+\phi) \end{bmatrix}$$

$$= \begin{bmatrix} \cos\phi & \sin\phi \\ -\sin\phi & \cos\phi \end{bmatrix} \quad (2.45)$$

Equating the elements gives the vector inner product and the exterior product (sloppily, the cross product without direction)[34]:

$$\cos\phi = \frac{xt+yz}{\sqrt{x^2+y^2}\sqrt{t^2+z^2}} \quad (2.46)$$

$$\sin\phi = \frac{xz-yt}{\sqrt{x^2+y^2}\sqrt{t^2+z^2}} \quad (2.47)$$

---

[34] The term exterior product is taken from the Clifford algebras and the Grassman algebras.

Note: The expression in 2-dimensional Euclidean vector algebra for the angle between two vectors, $\{\vec{a}=[x,y], \vec{b}=[t,z]\}$ is:

$$\cos\phi = \frac{\langle \vec{a}, \vec{b} \rangle}{|\vec{a}||\vec{b}|} = \frac{xt+yz}{\sqrt{x^2+y^2}\sqrt{t^2+z^2}} \tag{2.48}$$

The expression in 2-dimensional Euclidean vector algebra for the magnitude of the cross product of two vectors is:

$$\sin\phi = \frac{|\vec{a}\times\vec{b}|}{|\vec{a}||\vec{b}|} = \frac{xz-yt}{\sqrt{x^2+y^2}\sqrt{t^2+z^2}} \tag{2.49}$$

Thus the inner product and the exterior product have fallen out of the algebra of the Euclidean complex numbers as the way of calculating the angle between two vectors. We refer to the operation we have just shown as the angle product of two complex numbers[35], and we denote it with the symbol, $\odot$.

Definition: *The Angle Product*
The angle product of two matrices (within the same natural algebra) is the conjugate of one (either) matrix multiplied by the other matrix:

$$A \odot B = \overline{A}B \tag{2.50}$$

where the line atop the matrix denotes the conjugate of that matrix[36].

In the case of the Euclidean complex numbers, we have:

$$\begin{bmatrix} x & y \\ -y & x \end{bmatrix} \odot \begin{bmatrix} t & z \\ -z & t \end{bmatrix} = \begin{bmatrix} x & -y \\ y & x \end{bmatrix}\begin{bmatrix} t & z \\ -z & t \end{bmatrix} \tag{2.51}$$

$$= \begin{bmatrix} \begin{bmatrix} x \\ y \end{bmatrix} \cdot \begin{bmatrix} t \\ z \end{bmatrix} & \left|\begin{bmatrix} x \\ y \end{bmatrix} \wedge \begin{bmatrix} t \\ z \end{bmatrix}\right| \\ -\left|\begin{bmatrix} x \\ y \end{bmatrix} \wedge \begin{bmatrix} t \\ z \end{bmatrix}\right| & \begin{bmatrix} x \\ y \end{bmatrix} \cdot \begin{bmatrix} t \\ z \end{bmatrix} \end{bmatrix} \tag{2.52}$$

where we have used the wedge to indicate the exterior product and the modulus sign to indicate the magnitude of the wedge product. We note that:

---

[35] 2nd edition note: The term angle product has fallen out of use.
[36] Remember, the conjugate of an algebraic matrix form is the adjoint of that matrix form.

Section 2

$$\begin{bmatrix} a \\ b \end{bmatrix} \wedge \begin{bmatrix} c \\ d \end{bmatrix} = -\begin{bmatrix} c \\ d \end{bmatrix} \wedge \begin{bmatrix} a \\ b \end{bmatrix} \qquad (2.53)$$

The angle product of two vectors is a vector. We have:

$$[\phi] \odot [\theta] = \overline{[\theta] \odot [\phi]} \qquad (2.54)$$

$$(\alpha[\phi] + \beta[\lambda]) \odot [\theta] = \alpha[\phi] \odot [\theta] + \beta[\lambda] \odot [\theta] \qquad (2.55)$$

$$[\phi] \odot [\phi] = [\det[\phi]] > 0 \qquad (2.56)$$

Thus the angle product operation satisfies the axioms of an inner product[37], with the magnitude of the exterior product thrown in. The reader should note that the exterior product cannot be the cross product. The space we have found within the complex numbers is 2-dimensional Euclidean space; it has no third dimension[38].

It is generally understood (though kept secret from undergraduates) that the cross product operation does not produce a vector but instead produces a bi-vector. A bi-vector is seen by Clifford algebraists as an area (a 2-dimensional plane of finite extent). However, in 3-dimensional space, to every 2-dimensional plane there is a corresponding vector (perpendicular to the plane) up to direction (which is arbitrary). Thus, in 3-dimensional space, planes are dual to vectors.[39] It is this duality that allows us to mislead undergraduates into thinking that the output of the cross product operation is a vector. The particular vector depends upon the metric of the 3-dimensional space.

Summary:

The complex numbers are the algebraic vector product algebra of 2-dimensional Euclidean space. Within this algebra the algebraic vector product (matrix multiplication) creates the space. The metric and the inner product (and the wedge product) are simply calculative procedures within what is already a geometric space. This is very different from the

---

[37] If anyone is interested, we thus have a complete Hilbert space – which, in the case of the complex numbers, is nothing new.

[38] 2nd edition note: At this point, your author ought to have realised that rotation in the complex plane is not rotation about an axis. Instead, your author, in line with over 300 years of tradition, did not notice the absence of an axis. This absence of an axis of rotation is one of the properties of spinor rotations.

[39] This duality is known as the Hodge dual.

Complex Numbers The Higher Dimensional Forms – 2nd Edition

conventional mathematical procedure of endowing a space with structure by putting a metric and an inner product on to it.

In due course, we will see that all the natural algebras are algebraic vector product algebras, and that they are the only such algebras.

....

An Aside:
The Clifford Algebra of 2-dimensional Euclidean Space

Multiplicative distributivity is a central property of numbers. The inner product of two vectors is component-wise multiplication and is not distributive. If we multiply a vector by itself distributively, we get:

$$\left(a_1\vec{e}_1 + a_2\vec{e}_2\right)\left(a_1\vec{e}_1 + a_2\vec{e}_2\right) = a_1^2\vec{e}_1\vec{e}_1 + a_1a_2\vec{e}_1\vec{e}_2 + a_2a_1\vec{e}_2\vec{e}_1 + a_2^2\vec{e}_2\vec{e}_2$$

(2.57)

Setting this equal to the inner product, $a_1^2 + a_2^2$, (the cross product of a vector with itself is zero) compels us to set:

$$\left\{ |\vec{e}_1\vec{e}_1| = |\vec{e}_2\vec{e}_2| = 1, |\vec{e}_1\vec{e}_2| = -|\vec{e}_2\vec{e}_1| \right\}$$

(2.58)

Using these relations and multiplying two different vectors together distributively gives the $Cl_{2,0}$ [40] Clifford product:

$$\left(a_1\vec{e}_1 + a_2\vec{e}_2\right)\left(b_1\vec{e}_1 + b_2\vec{e}_2\right) = a_1b_1\vec{e}_1\vec{e}_1 + a_1b_2\vec{e}_1\vec{e}_2 + a_2b_1\vec{e}_2\vec{e}_1 + a_2b_2\vec{e}_2\vec{e}_2$$

$$= \left(a_1b_1 + a_2b_2\right) + \left(a_1b_2 - a_2b_1\right)\vec{e}_1\vec{e}_2$$

$$= \begin{bmatrix} a_1 \\ a_2 \end{bmatrix} \cdot \begin{bmatrix} b_1 \\ b_2 \end{bmatrix} + \begin{bmatrix} a_1 \\ a_2 \end{bmatrix} \wedge \begin{bmatrix} b_1 \\ b_2 \end{bmatrix}$$

(2.59)

Thus, the Clifford product produces both the inner product and the exterior product. The $\vec{e}_1\vec{e}_2$ object is a bi-vector. The Clifford product is not an algebraic multiplication because the product of two vectors is not a vector[41].

---

[40] $Cl_{2,0}$ is the name of the Clifford algebra of 2-dimensional Euclidean space.
[41] 2nd edition note: Clifford algebra is completely rewritten in "The Naked Spinor" by Dennis Morris

Section 2

Logarithms and Powers: [42]

The inverse of the exponential function of 2×2 matrices we call the logarithm function of 2×2 matrices. Like the exponential, it is a function that takes 2×2 matrices for its arguments. We denote it by $\log m_{[e]}$. Because it is the inverse of the exponential, then, if and only if A & B commute, we have:

$$\log m_{[e]}(AB) = \log m_{[e]}(A) + \log m_{[e]}(B) \qquad (2.60)$$

For the complex numbers, we have:

$$\begin{bmatrix} a & b \\ -b & a \end{bmatrix} = \begin{bmatrix} r & 0 \\ 0 & r \end{bmatrix} \begin{bmatrix} \cos\theta & \sin\theta \\ -\sin\theta & \cos\theta \end{bmatrix} \qquad (2.61)$$

And:

$$\log m_{[e]}\left(\begin{bmatrix} a & b \\ -b & a \end{bmatrix}\right)$$

$$= \log m_{[e]}\left(\begin{bmatrix} r & 0 \\ 0 & r \end{bmatrix} \begin{bmatrix} \cos\theta & \sin\theta \\ -\sin\theta & \cos\theta \end{bmatrix}\right) \qquad (2.62)$$

$$= \log m_{[e]}\left(\begin{bmatrix} r & 0 \\ 0 & r \end{bmatrix}\right) + \log m_{[e]}\left(\begin{bmatrix} \cos\theta & \sin\theta \\ -\sin\theta & \cos\theta \end{bmatrix}\right)$$

$$= \log m_{[e]}\left(\begin{bmatrix} r & 0 \\ 0 & r \end{bmatrix}\right) + \log m_{[e]}\left(\begin{bmatrix} \cos(\theta+2n\pi) & \sin(\theta+2n\pi) \\ -\sin(\theta+2n\pi) & \cos(\theta+2n\pi) \end{bmatrix}\right)$$

$$= \log m_{[e]}\left(\begin{bmatrix} r & 0 \\ 0 & r \end{bmatrix}\right) + \log m_{[e]}\left(\exp\left(\begin{bmatrix} 0 & (\theta+2n\pi) \\ -\theta+2n\pi & 0 \end{bmatrix}\right)\right) \qquad (2.63)$$

$$= \log m_{[e]}\left(\begin{bmatrix} r & 0 \\ 0 & r \end{bmatrix}\right) + \begin{bmatrix} 0 & \theta+2n\pi \\ -(\theta+2n\pi) & 0 \end{bmatrix} \qquad (2.64)$$

And so we have that the logarithm of a complex number is multi-valued.

For comparison, dropping from matrix notation into the usual notation:

---

[42] 2nd edition note: To date, logarithms and powers of these spinor algebras have found no use within physics.

Complex Numbers The Higher Dimensional Forms – 2nd Edition

$$\log m_{[e]}\left(\begin{bmatrix} r & 0 \\ 0 & r \end{bmatrix}\right) + \log m_{[e]}\left(\begin{bmatrix} \cos\theta & \sin\theta \\ -\sin\theta & \cos\theta \end{bmatrix}\right)$$

$$\equiv \log_e r + \log_e \left(\cos\theta + \hat{i}\sin\theta\right)$$

$$\equiv \log_e r + \log_e \left(\frac{e^{\hat{i}\theta} + e^{-\hat{i}\theta}}{2} + \hat{i}\frac{e^{\hat{i}\theta} - e^{-\hat{i}\theta}}{2\hat{i}}\right) \qquad (2.65)$$

$$\equiv \log_e r + \log_e \left(e^{\hat{i}\theta}\right)$$

$$\equiv \log_e r + \log_e \left(e^{\hat{i}(\theta + 2n\pi)}\right)$$

$$\equiv \log_e r + \hat{i}(\theta + 2n\pi)$$

With which the reader might be more familiar.

We can use the logarithm to calculate exponentials. For example, calculate:

$$\begin{bmatrix} 0 & b \\ -b & 0 \end{bmatrix}^{\begin{bmatrix} 0 & 1 \\ -1 & 0 \end{bmatrix}} \qquad (2.66)$$

We set:

$$\begin{bmatrix} 0 & b \\ -b & 0 \end{bmatrix} = \exp\left(\log m_{[e]}\left(\begin{bmatrix} 0 & b \\ -b & 0 \end{bmatrix}\right)\right) \qquad (2.67)$$

$$\begin{bmatrix} 0 & b \\ -b & 0 \end{bmatrix}^{\begin{bmatrix} 0 & 1 \\ -1 & 0 \end{bmatrix}} = \exp\left(\log m_{[e]}\left(\begin{bmatrix} 0 & b \\ -b & 0 \end{bmatrix}\right)\right)^{\begin{bmatrix} 0 & 1 \\ -1 & 0 \end{bmatrix}}$$

$$= \exp\left(\begin{bmatrix} 0 & 1 \\ -1 & 0 \end{bmatrix}\log m_{[e]}\left(\begin{bmatrix} 0 & b \\ -b & 0 \end{bmatrix}\right)\right) \qquad (2.68)$$

The equations $\{0 = r\cos\theta, b = r\sin\theta\}$ have solutions

$$\left\{\left\{r = b, \theta = \frac{\pi}{2}\right\}, \left\{r = -b, \theta = -\frac{\pi}{2}\right\}\right\}.$$

We take the positive solution:

Section 2

$$\begin{bmatrix} 0 & b \\ -b & 0 \end{bmatrix}^{\begin{bmatrix} 0 & 1 \\ -1 & 0 \end{bmatrix}}$$

$$= \exp\left(\begin{bmatrix} 0 & 1 \\ -1 & 0 \end{bmatrix} \log m_{[,]} \left(\begin{bmatrix} b & 0 \\ 0 & b \end{bmatrix} \right) + \begin{bmatrix} 0 & \frac{\pi}{2} + 2n\pi \\ -\left(\frac{\pi}{2} + 2n\pi\right) & 0 \end{bmatrix} \right)$$

$$= \exp\left(\begin{bmatrix} 0 & 1 \\ -1 & 0 \end{bmatrix} \log m_{[,]} \left(\begin{bmatrix} b & 0 \\ 0 & b \end{bmatrix} \right) \right) \exp\left(\begin{bmatrix} 0 & 1 \\ -1 & 0 \end{bmatrix} \begin{bmatrix} 0 & \frac{\pi}{2} + 2n\pi \\ -\left(\frac{\pi}{2} + 2n\pi\right) & 0 \end{bmatrix} \right)$$

$$= \exp\left(\begin{bmatrix} 0 & \log_r b \\ -\log_r b & 0 \end{bmatrix} \right) \exp\left(\begin{bmatrix} -\left(\frac{\pi}{2} + 2n\pi\right) & 0 \\ 0 & -\left(\frac{\pi}{2} + 2n\pi\right) \end{bmatrix} \right)$$

$$= \begin{bmatrix} \cos(\log_r b) & \sin(\log_r b) \\ -\sin(\log_r b) & \cos(\log_r b) \end{bmatrix} \exp\left(\begin{bmatrix} -\left(\frac{\pi}{2} + 2n\pi\right) & 0 \\ 0 & -\left(\frac{\pi}{2} + 2n\pi\right) \end{bmatrix} \right)$$

(2.69)

If $b = 1$, this is:

$$\begin{bmatrix} 0 & 1 \\ -1 & 0 \end{bmatrix}^{\begin{bmatrix} 0 & 1 \\ -1 & 0 \end{bmatrix}} = \begin{bmatrix} 1 & 0 \\ 0 & 1 \end{bmatrix} \exp\left(\begin{bmatrix} -\left(\frac{\pi}{2} + 2n\pi\right) & 0 \\ 0 & -\left(\frac{\pi}{2} + 2n\pi\right) \end{bmatrix} \right) \quad (2.70)$$

Which, written in the usual complex notation, is the familiar:

$$\hat{i}^{\hat{i}} = e^{-\frac{\pi}{2} + 2n\pi} \quad (2.71)$$

We note that the exponential is multi-valued.

34

Vector calculus in 2-dimensional space:

A function upon a space is a scalar field upon that space. In matrix form, this is:

$$\begin{bmatrix} f(a,b) & 0 \\ 0 & f(a,b) \end{bmatrix} \tag{2.72}$$

The differential is formed by taking the partial derivatives. The matrix configuration identifies these partial derivatives as components of a vector, and so we do not need to keep them separate:

$$d\begin{bmatrix} f(a,b) & 0 \\ 0 & f(a,b) \end{bmatrix} = \frac{\partial \begin{bmatrix} f & 0 \\ 0 & f \end{bmatrix}}{\partial \begin{bmatrix} a & 0 \\ 0 & a \end{bmatrix}} + \frac{\partial \begin{bmatrix} f & 0 \\ 0 & f \end{bmatrix}}{\partial \begin{bmatrix} 0 & b \\ -b & 0 \end{bmatrix}}$$

$$= \frac{\partial \begin{bmatrix} f & 0 \\ 0 & f \end{bmatrix}}{\partial \begin{bmatrix} a & 0 \\ 0 & a \end{bmatrix}} \begin{bmatrix} 1 & 0 \\ 0 & 1 \end{bmatrix} + \frac{\partial \begin{bmatrix} f & 0 \\ 0 & f \end{bmatrix}}{\partial \begin{bmatrix} b & 0 \\ 0 & b \end{bmatrix}} \begin{bmatrix} 0 & 1 \\ -1 & 0 \end{bmatrix} \tag{2.73}$$

$$= \begin{bmatrix} \dfrac{\partial f}{\partial a} & -\dfrac{\partial f}{\partial b} \\ \dfrac{\partial f}{\partial b} & \dfrac{\partial f}{\partial a} \end{bmatrix}$$

Of course, the differential is the gradient of a scalar field, $grad(f)$ which is a vector[43].

The differential of a vector field is:

$$d\begin{bmatrix} F_a(a,b) & F_b(a,b) \\ -F_b(a,b) & F_a(a,b) \end{bmatrix} = \frac{\partial \begin{bmatrix} F_a & F_b \\ -F_b & F_a \end{bmatrix}}{\partial \begin{bmatrix} a & 0 \\ 0 & a \end{bmatrix}} + \frac{\partial \begin{bmatrix} F_a & F_b \\ -F_b & F_a \end{bmatrix}}{\partial \begin{bmatrix} b & 0 \\ 0 & b \end{bmatrix}} \begin{bmatrix} 0 & 1 \\ -1 & 0 \end{bmatrix}$$

(2.74)

---

[43] Technically, the gradient is a 1-form. We have a 1-form – look at the position of the minus sign.

Section 2

$$= \begin{bmatrix} \dfrac{\partial F_a}{\partial a} + \dfrac{\partial F_b}{\partial b} & \dfrac{\partial F_b}{\partial a} - \dfrac{\partial F_a}{\partial b} \\ -\left(\dfrac{\partial F_b}{\partial a} - \dfrac{\partial F_a}{\partial b}\right) & \dfrac{\partial F_a}{\partial a} + \dfrac{\partial F_b}{\partial b} \end{bmatrix}$$

$$= \begin{bmatrix} div(F) & curl(F) \\ -curl(F) & div(F) \end{bmatrix}$$

(2.75)

We get the Laplacian of a scalar field and of a vector field by differentiating yet again.

Section 3

Introduction to Section 3:

In this section, we seek to further familiarise the reader with doing algebra in matrix notation and to introduce the reader to another natural algebra – another form of complex numbers. We seek to prepare the reader for what comes after – an infinite number of new types of complex numbers each containing a new type of space. This section is the final part of the gentle introduction that we have given the reader. We hope that it will open the reader's mind into a state of readiness to receive new ideas and extend old ones.

The Hyperbolic Complex Numbers – Basics:

Readers will be familiar with the Euclidean complex numbers but might be wholly unfamiliar with the hyperbolic complex numbers. Hyperbolic complex numbers are also called Study numbers[44]. It might be, and your author is far from certain about this, that Study numbers are named after Eduard Study, a German mathematician working around the year 1900[45]. Your author has scanned through Eduard Study's works but has found nothing at all concerning these hyperbolic complex numbers,[46] and your author wonders if he is erroneously attributing them to Eduard Study. The fancy 'S' denoting them is an invention of your author. Having placed the credit where credit might be due, we will henceforward refer to this algebra as the hyperbolic complex numbers.

> 2nd edition note:
> It is now known that the hyperbolic complex numbers have been independently discovered many times. The earliest known discovery is by Cockle in 1848, Cockle was president of the London Mathematical Society from 1886 to 1888, and so it is remarkable that these numbers are so little known. These numbers have been given many different names by different discoverers. We have:

---

[44] Your author discovered the name Study numbers in the book 'Clifford Algebras and Spinors' (page 24) by Pertti Lounesto - ISBN 0-521-00551-5. This is the only instance I know of them being mentioned anywhere, and only four sentences. Unfortunately, Pertti Lounesto has died.

[45] Readers seeking his works are directed to the web site *copac.ac.uk*.

[46] This might be because, although the mathematics is written in mathematics, the English is written in German in which your author is not fluent.

Tessarines	Cockle	1848
Algebraic Motors	Clifford	1882
Hyperbolic complex numbers	Vignaux	1935
Double numbers	Yaglorn	1968
Anormal complex numbers	Benz	1973
Perplex numbers	Fjelstad	1986
Double numbers	Hazelwinkle	1990
Lorentz numbers	Harvey	1990
Dual numbers	Hucks	1993
Hyperbolic complex numbers	Sobczyk	1995
Split complex numbers	Rosenfield	1997
Study numbers	Lounesto	2001

The hyperbolic complex numbers are a multiplicatively commutative division algebra; they are an algebraic field. Thus, their algebraic status, with the restricted values, is equal to that of the complex numbers.[47]

The matrix representation of a hyperbolic complex number, $\left(a + \hat{r}b\right) : \hat{r} = \sqrt{+1}$, is:

$$\begin{bmatrix} a & b \\ b & a \end{bmatrix} : |a| > |b| \tag{3.1}$$

in which, as always within this work, the matrix elements are real numbers. The reader might be inclined to think $\hat{r} = -1$. However, the matrices:

$$\left\{ \begin{bmatrix} 0 & 1 \\ 1 & 0 \end{bmatrix}, \begin{bmatrix} -1 & 0 \\ 0 & -1 \end{bmatrix} \right\} \tag{3.2}$$

are not the same. The restrictions are imposed to guarantee non-singularity of the matrix. In practice, we achieve non-singularity by forming the polar form of the algebra by taking the exponential of the matrix (3.1).

Addition is matrix addition, and multiplication is matrix multiplication. The conjugate is the adjoint matrix:

$$\begin{bmatrix} a & -b \\ -b & a \end{bmatrix} \tag{3.3}$$

In non-matrix terms, hyperbolic complex numbers are similar to complex numbers, but instead of having $\hat{i} = \sqrt{-1}$ they have $\hat{r} = \sqrt{+1}$. They have

---

[47] The hyperbolic complex numbers are the even sub-algebra, $Cl_{1,1}^+$, of the Clifford algebra of $\mathbb{R}^{1,1}$ which is $Cl_{1,1}$.

Cartesian form $x+\hat{r}y : |x|>|y|$. Addition and multiplication are parallels of the complex numbers:

$$(a+\hat{r}b)+(x+\hat{r}y) = (a+x)+\hat{r}(b+y) \tag{3.4}$$

$$(a+\hat{r}b)(x+\hat{r}y) = (ax+\hat{r}^2 by)+\hat{r}(ay+bx)$$
$$= (ax+by)+\hat{r}(ay+bx) \tag{3.5}$$

The conjugate of the hyperbolic complex number $x+\hat{r}y$ is $x-\hat{r}y$, and division is done with the conjugate as with complex numbers:

$$\frac{a+\hat{r}b}{x+\hat{r}y} = \frac{(a+\hat{r}b)(x-\hat{r}y)}{(x+\hat{r}y)(x-\hat{r}y)} = \frac{(a+\hat{r}b)(x-\hat{r}y)}{x^2-y^2} \tag{3.6}$$

Such division is not defined when $|x|=|y|$, but these values are excluded values.

Hyperbolic complex numbers sit on a plane in a way similar to the way complex numbers sit on the Argand diagram, but the hyperbolic complex number plane is a hyperbolic space rather than a Euclidean one. As visualized by our Euclidean eyes, the hyperbolic complex number plane is split into two halves[48] by the 45° lines, where $|x|=|y|$, and the hyperbolic complex numbers occupy two horizontally opposite quadrants. The real numbers lie on the horizontal x-axis. Below is an Argand diagram for hyperbolic complex numbers. It is a plot of hyperbolas with $h = 1, 2, 3, 4$. Also plotted are the 45° & minus 45° axes. The hyperbolas are the 'real' co-ordinates. Thus, for example, all hyperbolic complex numbers with the real part, say, $a = 3$ are situated upon the hyperbola that passes through the number 3 on the horizontal axis.

---

[48] 2nd edition note: We now know that the hyperbolic complex numbers exist only in one half. The part with the negative real axis ought to have been excluded. Your author included both halves to satisfy the standard axioms of an algebraic field. Your author ought to have ignored these man made axioms and preferred the simple fact that we cannot travel backwards in time. The hyperbolic complex numbers are the space-time of special relativity; see: Dennis Morris Empty Space is Amazing Stuff. ISBN: 978-0954978075 Pantaneto Press.

Section 3

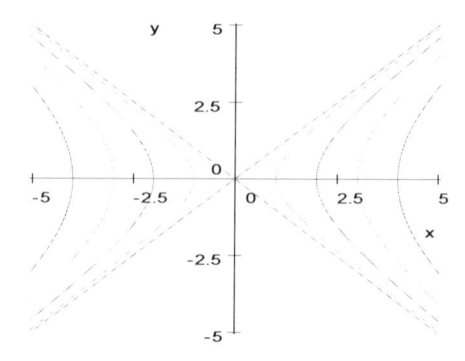

Argand diagram for the hyperbolic complex numbers

The vertically opposite quadrants are not part of the hyperbolic complex number space; they are the realm of numbers of the Cartesian form $a + \hat{r}b : |a| < |b|$. These numbers are not an algebra.

Addendum:

Proof that the two Notations of Hyperbolic Complex Numbers are Algebraically Isomorphic.

Theorem

The algebra of matrices of the form $\begin{bmatrix} a & b \\ b & a \end{bmatrix} : |a| > |b|$ is isomorphic to the algebra of the hyperbolic complex numbers, $a + \hat{r}b : |a| > |b|$.

Proof:

$$\phi : (\mathbb{S}, +, \times) \mapsto (M_{(2,\mathbb{R})}, +, \bullet) \quad : \quad \phi(a + \hat{r}b) \mapsto \begin{bmatrix} a & b \\ b & a \end{bmatrix}$$

$$\phi\left((a + \hat{r}b) + (e + \hat{r}f)\right) \mapsto \begin{bmatrix} a+e & b+f \\ b+f & a+e \end{bmatrix} = \begin{bmatrix} a & b \\ b & a \end{bmatrix} + \begin{bmatrix} e & f \\ f & e \end{bmatrix}$$

$$\phi\left((a + \hat{r}b) \times (e + \hat{r}f)\right) \mapsto \begin{bmatrix} ae+bf & af+be \\ af+be & ae+bf \end{bmatrix} = \begin{bmatrix} a & b \\ b & a \end{bmatrix} \bullet \begin{bmatrix} e & f \\ f & e \end{bmatrix}$$

$$\phi\left(\alpha(a + \hat{r}b)\right) \mapsto \begin{bmatrix} \alpha a & \alpha b \\ \alpha b & \alpha a \end{bmatrix} = \alpha \begin{bmatrix} a & b \\ b & a \end{bmatrix}$$

The C-R Eqns. for Hyperbolic Complex Numbers:

We consider functions of hyperbolic complex numbers. These are of the form:

$$f\left(\begin{bmatrix} a & b \\ b & a \end{bmatrix}\right) \to \begin{bmatrix} u(a,b) & v(a,b) \\ v(a,b) & u(a,b) \end{bmatrix} \quad (3.7)$$

where we require $|u(a,b)| > |v(a,b)|$ for the allowed values of a & b.

Differentiation leads to the Cauchy-Riemann equations of the hyperbolic complex numbers. The matrix function is differentiable only if:

$$\frac{\partial u}{\partial a} = \frac{\partial v}{\partial b} \quad \& \quad \frac{\partial u}{\partial b} = \frac{\partial v}{\partial a} \quad (3.8)$$

These are the Cauchy-Riemann equations of the hyperbolic complex numbers.

The Polar Form of Hyperbolic Complex Numbers:

Consider:

$$\exp\left(\begin{bmatrix} a & b \\ b & a \end{bmatrix}\right) = \begin{bmatrix} e^a & 0 \\ 0 & e^a \end{bmatrix} \begin{bmatrix} 1 + \frac{b^2}{2!} + \frac{b^4}{4!} + \ldots & b + \frac{b^3}{3!} + \frac{b^5}{5!} + \ldots \\ b + \frac{b^3}{3!} + \frac{b^5}{5!} + \ldots & 1 + \frac{b^2}{2} + \frac{b^4}{4!} + \ldots \end{bmatrix} \quad (3.9)$$

$$= \begin{bmatrix} e^a & 0 \\ 0 & e^a \end{bmatrix} \begin{bmatrix} \cosh b & \sinh b \\ \sinh b & \cosh b \end{bmatrix}$$

$$= \begin{bmatrix} r & 0 \\ 0 & r \end{bmatrix} \begin{bmatrix} \cosh b & \sinh b \\ \sinh b & \cosh b \end{bmatrix} \quad (3.10)$$

We have a length matrix and the rotation matrix of 2-dimensional hyperbolic space. The trace of the b-sub-matrix is zero, and so the determinant of the rotation matrix is unity, and we have the identity: $\cosh^2 b - \sinh^2 b = 1$.

We have:

$$\begin{bmatrix} x & y \\ y & x \end{bmatrix} = \begin{bmatrix} r & 0 \\ 0 & r \end{bmatrix} \begin{bmatrix} \cosh b & \sinh b \\ \sinh b & \cosh b \end{bmatrix} \quad (3.11)$$

$$\det\left(\begin{bmatrix} x & y \\ y & x \end{bmatrix}\right) = \det\left(\begin{bmatrix} r & 0 \\ 0 & r \end{bmatrix} \begin{bmatrix} \cosh b & \sinh b \\ \sinh b & \cosh b \end{bmatrix}\right) \quad (3.12)$$

$$x^2 - y^2 = r^2$$

Leading to the distance function[49] of 2-dimensional hyperbolic space:
$$r = \sqrt{x^2 - y^2} \qquad (3.13)$$
We note that the polar form can never have determinant equal to zero except for the zero matrix. The polar form does not need to be restricted to guarantee non-singularity. The reader should note that $|\cosh \chi| > |\sinh \chi| \; \forall \chi$.

Important Point: We do not need to restrict the algebra of hyperbolic complex numbers in polar form. This means that the algebra consists of the unrestricted matrices of the form:
$$\begin{bmatrix} h & 0 \\ 0 & h \end{bmatrix} \begin{bmatrix} \cosh \chi & \sinh \chi \\ \sinh \chi & \cosh \chi \end{bmatrix} \qquad (3.14)$$

We make the same points that we made when we found the polar form of the Euclidean complex numbers in Section 2.
1) The trigonometric functions associated with 2-dimensional hyperbolic space fall out of the algebra of the hyperbolic complex numbers.
2) The rotation matrix of 2-dimensional hyperbolic space falls out of the algebra of the hyperbolic complex numbers.
3) The distance function of 2-dimensional hyperbolic space falls out of the algebra of the hyperbolic complex numbers.

We take the view that this is sufficient to say that 2-dimensional hyperbolic space is contained within the algebra of the hyperbolic complex numbers.

Theorem:
2-dimensional hyperbolic space is contained within the algebra of the hyperbolic complex numbers.
Proof: See Above

With this theorem, we have effectively unified the geometry of 2-dimensional hyperbolic space and the algebra of the hyperbolic complex numbers.

In contradistinction to the Euclidean complex numbers, the distance function is anti-symmetric in $\{x, y\}$, by which we mean that, if $\{x, y\}$ are swapped around, the sign of the output of the distance function is changed. In particular, we point out that the x-axis is of a different nature as the y-axis; if

---

[49] Technically, this function is not a metric – it violates the triangle inequality, axiom 3. See Appendix 1.

one is a spatial axis, the other is not; if one is a temporal axis, the other is not.

2-dimensional hyperbolic space (otherwise known as Minkowski space-time) is the space-time of special relativity. Special relativity is a 2-dimensional theory, and the space-time 'metric' often given, $\tau^2 = t^2 - x^2 - y^2 - z^2$, is incorrect[50]. The correct form is: $\tau^2 = t^2 - x^2$. The extra terms are added in a 'fudge' designed to take account of the three spatial dimensions we perceive.

The hyperbolic complex numbers have polar form based on the hyperbolic trigonometric functions *cosh* & *sinh*. These are not periodic functions. The Euclidean complex numbers have polar form based on the Euclidean trigonometric functions *cos* & *sin*. These are periodic functions. It is the periodic nature of the Euclidean trigonometric functions that 'causes' the multi-valued nature of many complex functions. The non-periodic nature of the hyperbolic trigonometric functions leads to the non-periodic nature of functions within this algebra.

Logarithms and Powers:

The Euler relations for the hyperbolic trigonometric functions are usually given as:

$$\cosh \chi = \frac{e^\chi + e^{-\chi}}{2}, \quad \sinh \chi = \frac{e^\chi - e^{-\chi}}{2} \tag{3.15}$$

We consider:

$$e^{\hat{r}\chi} = 1 + \hat{r}\chi + \frac{\hat{r}^2 \chi^2}{2!} + \frac{\hat{r}^3 \chi^3}{3!} + ... \tag{3.16}$$

where $\hat{r} = \sqrt{+1} \neq \pm 1$. This leads to:

$$e^{\hat{r}\chi} = \cosh \chi + \hat{r} \sinh \chi \tag{3.17}$$

Similarly:

$$e^{-\hat{r}\chi} = 1 - \hat{r}\chi + \frac{\hat{r}^2 \chi^2}{2!} - \frac{\hat{r}^3 \chi^3}{3!} + ... \tag{3.18}$$

This leads to:

$$e^{-\hat{r}\chi} = \cosh \chi - \hat{r} \sinh \chi \tag{3.19}$$

---

[50] 2nd edition note: 'incorrect' for special relativity but correct for general relativity. SR exists in a division algebra space (spinor space). GR exists in an emergent expectation space. See: Dennis Morris Upon General Relativity.

Section 3

Combining these results gives:

$$\cosh \hat{\chi} = \frac{e^{\hat{r}\chi} + e^{-\hat{r}\chi}}{2}$$

$$\sinh \hat{\chi} = \frac{e^{\hat{r}\chi} - e^{-\hat{r}\chi}}{2\hat{r}}$$

(3.20)

The reader might like to compare these to the Euclidean Euler relations:

$$\cos \theta = \frac{e^{\hat{i}\theta} + e^{-\hat{i}\theta}}{2}, \quad \sin \theta = \frac{e^{\hat{i}\theta} - e^{-\hat{i}\theta}}{2\hat{i}} \tag{3.21}$$

No more need be said.

For the hyperbolic complex numbers, we have:

$$\begin{bmatrix} a & b \\ b & a \end{bmatrix} = \begin{bmatrix} h & 0 \\ 0 & h \end{bmatrix} \begin{bmatrix} \cosh \chi & \sinh \chi \\ \sinh \chi & \cosh \chi \end{bmatrix} \tag{3.22}$$

$$\log m_{[e]}\left(\begin{bmatrix} a & b \\ b & a \end{bmatrix}\right) = \log m_{[e]}\left(\begin{bmatrix} h & 0 \\ 0 & h \end{bmatrix}\begin{bmatrix} \cosh \chi & \sinh \chi \\ \sinh \chi & \cosh \chi \end{bmatrix}\right)$$

$$= \log m_{[e]}\left(\begin{bmatrix} h & 0 \\ 0 & h \end{bmatrix}\right) + \log m_{[e]}\left(\begin{bmatrix} \cosh \chi & \sinh \chi \\ \sinh \chi & \cosh \chi \end{bmatrix}\right)$$

$$= \log m_{[e]}\left(\begin{bmatrix} h & 0 \\ 0 & h \end{bmatrix}\right) + \log m_{[e]}\left(\exp\left(\begin{bmatrix} 0 & \chi \\ \chi & 0 \end{bmatrix}\right)\right)$$

$$= \log m_{[e]}\left(\begin{bmatrix} h & 0 \\ 0 & h \end{bmatrix}\right) + \begin{bmatrix} 0 & \chi \\ \chi & 0 \end{bmatrix}$$

(3.23)

And, so, we have that the logarithm of a hyperbolic complex number is not multi-valued.

Theorem:
Logarithms of Hyperbolic Complex Numbers are not multi-valued.
Proof: See Above

Dropping from matrix notation into non-matrix notation:

$$\log m_{[e]}\left(\begin{bmatrix} h & 0 \\ 0 & h \end{bmatrix}\right) + \log m_{[e]}\left(\begin{bmatrix} \cosh \chi & \sinh \chi \\ \sinh \chi & \cosh \chi \end{bmatrix}\right)$$

$$= \log_e h + \log_e \left(\cosh \chi + \hat{r} \sinh \chi\right)$$

(3.24)

$$\equiv \log_e h + \log_e \left(\frac{e^{\hat{r}x} + e^{-\hat{r}x}}{2} + \hat{r} \frac{e^{\hat{r}x} - e^{-\hat{r}x}}{2\hat{r}} \right)$$

$$\equiv \log_e h + \log_e \left(e^{\hat{r}x} \right) \qquad (3.25)$$

$$\equiv \log_e h + \hat{r}\chi$$

Odd Parity Matrices, Quaternions, and Dihedralions:

In this sub-section, we digress from our path a little. The complex numbers are a sub-algebra of the quaternions. As a means of comparison, we consider the hyperbolic complex numbers in this light.

Of all the 2×2 matrices with characterizations:
$$\{a_{11} \in \mathbb{R}, a_{21} \in \mathbb{R}, a_{21} = \pm a_{12}, a_{22} = \pm a_{11}\} \qquad (3.26)$$
there is only one matrix with zero minus signs. That one is:
$$\mathbb{S} \simeq \begin{bmatrix} a & b \\ b & a \end{bmatrix} \triangleq \begin{bmatrix} -a & -b \\ -b & -a \end{bmatrix} \triangleq \begin{bmatrix} -a & b \\ b & -a \end{bmatrix} \triangleq \begin{bmatrix} a & -b \\ -b & a \end{bmatrix} \qquad (3.27)$$
where we have used the $\triangleq$ sign to indicate equality of matrix form. We have found that the algebra of these matrices, when restricted to $|a| > |b|$, is the algebra of the hyperbolic complex numbers. There are two 2×2 such matrices with one minus quantity. Those two are:
$$\mathbb{C} \simeq \begin{bmatrix} a & -b \\ b & a \end{bmatrix} \triangleq \begin{bmatrix} a & b \\ -b & a \end{bmatrix} \triangleq \begin{bmatrix} -a & -b \\ b & -a \end{bmatrix} \triangleq \begin{bmatrix} -a & b \\ -b & -a \end{bmatrix} \qquad (3.28)$$

$$\mathbb{OP}_\mathbb{C} = \begin{bmatrix} a & b \\ b & -a \end{bmatrix} \triangleq \begin{bmatrix} -a & b \\ b & a \end{bmatrix} \triangleq \begin{bmatrix} a & -b \\ -b & -a \end{bmatrix} \triangleq \begin{bmatrix} -a & -b \\ -b & a \end{bmatrix} \qquad (3.29)$$

The first of these is the complex numbers. To the second we have given the appellation $\mathbb{OP}_\mathbb{C}$[51]. There is only one matrix with two minus quantities. That one is:
$$\mathbb{OP}_\mathbb{S} = \begin{bmatrix} a & -b \\ b & -a \end{bmatrix} \triangleq \begin{bmatrix} -a & b \\ -b & a \end{bmatrix} \triangleq \begin{bmatrix} -a & -b \\ b & a \end{bmatrix} \triangleq \begin{bmatrix} a & b \\ -b & -a \end{bmatrix} \qquad (3.30)$$

To this matrix form we have given the appellation $\mathbb{OP}_\mathbb{S}$. The matrices with three or four minus quantities are included in the above.

---

[51] Odd Parity

Section 3

Let us first consider the third of these matrix types:

$$\mathbb{OP}_\mathbb{C} = \begin{bmatrix} a & b \\ b & -a \end{bmatrix} \qquad (3.31)$$

When two of these matrices are multiplied together they form a complex number. When three of these matrices are multiplied together they return to their own form. When four of these matrices are multiplied together they form a complex number; and so it goes on:

$$\begin{bmatrix} a & b \\ b & -a \end{bmatrix}\begin{bmatrix} e & f \\ f & -e \end{bmatrix} = \begin{bmatrix} ae+bf & af-be \\ -(af-be) & ae+bf \end{bmatrix} \qquad (3.32)$$

$$\begin{bmatrix} ae+bf & af-be \\ -(af-be) & ae+bf \end{bmatrix}\begin{bmatrix} m & n \\ n & -m \end{bmatrix} =$$

$$\begin{bmatrix} m(ae+bf)+n(af-be) & n(ae+bf)-m(af-be) \\ n(ae+bf)-m(af-be) & -(m(ae+bf)+n(af-be)) \end{bmatrix} \qquad (3.33)$$

These matrices are not multiplicatively commutative, but, if the multiplying order is reversed, they multiply to the conjugate of the complex number:

$$\begin{bmatrix} e & f \\ f & -e \end{bmatrix}\begin{bmatrix} a & b \\ b & -a \end{bmatrix} = \begin{bmatrix} ae+bf & -(af-be) \\ af-be & ae+bf \end{bmatrix} \qquad (3.34)$$

They square to a real number:

$$\begin{bmatrix} a & b \\ b & -a \end{bmatrix}\begin{bmatrix} a & b \\ b & -a \end{bmatrix} = \begin{bmatrix} a^2+b^2 & 0 \\ 0 & a^2+b^2 \end{bmatrix} \qquad (3.35)$$

What kind of a creature is it whose product is a complex number and whose square is a real number[52]. It is comprised of two 'bits', the a 'bit' and the b 'bit'. Thus, it is of the form: $a\hat{j} + b\hat{k}$. If we square such a creature, we get a positive real number, $a^2 + b^2$:

$$a^2\hat{j}^2 + b^2\hat{k}^2 + ab\left(\hat{j}\hat{k} + \hat{k}\hat{j}\right) = a^2 + b^2 \qquad (3.36)$$

Therefore, we conclude that $\hat{j}\hat{k} = -\hat{k}\hat{j}$ and that $\hat{j}^2 = +1$ and that $\hat{k}^2 = +1$.

Multiplied together two different such creatures are:

$$\left(a\hat{j} + b\hat{k}\right)\left(e\hat{j} + f\hat{k}\right) = ae\hat{j}^2 + bf\hat{k}^2 + af\hat{j}\hat{k} + be\hat{k}\hat{j} \qquad (3.37)$$

This, we know, is a complex number. The imaginary part is:

$$af\hat{j}\hat{k} + be\hat{k}\hat{j} = iy \qquad (3.38)$$

---

[52] 2nd edition note: It is a creature from the non-commutative algebras of the commutative $C_2 \times C_2$ group. These algebras were not known when the 1st edition of this book was written.

Using the above, we conclude that $\hat{j}\hat{k} = \pm i$ and $\hat{k}\hat{j} = \mp i$.

Some Clifford Algebra

The Clifford algebra $Cl_{2,0}$ is a 4-dimensional algebra with basis:

$$\{1, \vec{e}_1, \vec{e}_2, \vec{e}_1\vec{e}_2\}$$

These basis elements are algebraically isomorphic to the matrices:

$$\left\{ 1 \cong \begin{bmatrix} 1 & 0 \\ 0 & 1 \end{bmatrix}, \vec{e}_1 \cong \begin{bmatrix} 1 & 0 \\ 0 & -1 \end{bmatrix}, \vec{e}_2 \cong \begin{bmatrix} 0 & 1 \\ 1 & 0 \end{bmatrix}, \vec{e}_1\vec{e}_2 \cong \begin{bmatrix} 0 & 1 \\ -1 & 0 \end{bmatrix} \right\} \quad (3.39)$$

We have:

$$\mathbb{OP}_c = \begin{bmatrix} a & b \\ b & -a \end{bmatrix} \cong a\vec{e}_1 + b\vec{e}_2$$

$$\mathbb{C} = \begin{bmatrix} a & b \\ -b & a \end{bmatrix} \cong a + b\vec{e}_1\vec{e}_2 \quad (3.40)$$

2nd edition note: The whole of Clifford algebra, and with it spinor theory, in the book "The Naked Spinor" by Dennis Morris.

Let us now consider the algebra of the fourth of these matrix types:

$$\mathbb{OP}_s = \begin{bmatrix} a & -b \\ b & -a \end{bmatrix} \quad (3.41)$$

We shall consider only the matrices of this form that have $|a^2| > |b^2|$. When two of these matrices are multiplied together they form a hyperbolic complex number. When three of these matrices are multiplied together they return to their own form; and so it goes on:

$$\begin{bmatrix} a & -b \\ b & -a \end{bmatrix} \begin{bmatrix} e & -f \\ f & -e \end{bmatrix} = \begin{bmatrix} ae - bf & -(af - be) \\ -(af - be) & ae - bf \end{bmatrix} \quad (3.42)$$

$$\begin{bmatrix} ae - bf & -(af - be) \\ -(af - be) & ae - bf \end{bmatrix} \begin{bmatrix} m & -n \\ n & -m \end{bmatrix} =$$

$$\begin{bmatrix} m(ae - bf) - n(af - be) & -(n(ae - bf) - m(af - be)) \\ n(ae - bf) - m(af - be) & -(m(ae - bf) - n(af - be)) \end{bmatrix} \quad (3.43)$$

These matrices are not multiplicatively commutative, but, if the multiplying order is reversed, they multiply to the conjugate of the hyperbolic complex number:

47

Section 3

$$\begin{bmatrix} e & -f \\ f & -e \end{bmatrix} \begin{bmatrix} a & -b \\ b & -a \end{bmatrix} = \begin{bmatrix} ae+bf & af-be \\ af-be & ae+bf \end{bmatrix} \quad (3.44)$$

They square to a positive real number:

$$\begin{bmatrix} a & -b \\ b & -a \end{bmatrix} \begin{bmatrix} a & -b \\ b & -a \end{bmatrix} = \begin{bmatrix} a^2-b^2 & 0 \\ 0 & a^2-b^2 \end{bmatrix} \quad (3.45)$$

What kind of a creature is it whose product is a hyperbolic complex number and whose square is a real number. It is comprised of two 'bits', the a 'bit' and the b 'bit'. Thus, it is of the form: $a\hat{s}+b\hat{t}$. If we square such a creature, we get a real number, a^2-b^2:

$$a^2\hat{s}^2 + b^2\hat{t}^2 + ab\left(\hat{s}\hat{t}+\hat{t}\hat{s}\right) = a^2-b^2 \quad (3.46)$$

Therefore, we conclude that $\hat{s}\hat{t} = -\hat{t}\hat{s}$ and that $\hat{s}^2 = +1$ and that $\hat{t}^2 = -1$. Multiplied together two different such creatures are:

$$\left(a\hat{s}+b\hat{t}\right)\left(e\hat{s}+f\hat{t}\right) = ae\hat{s}^2 + bf\hat{t}^2 + af\hat{s}\hat{t} + be\hat{t}\hat{s} \quad (3.47)$$

This, we know, is a hyperbolic complex number. The 'imaginary' part is:

$$af\hat{s}\hat{t} + be\hat{t}\hat{s} = r\hat{y} \quad (3.48)$$

where we take $\hat{r} = \sqrt{+1}$

Using the above, we conclude that $\hat{s}\hat{t} = \pm\hat{r}$ and $\hat{t}\hat{s} = \mp\hat{r}$.

More Clifford Algebra

The Clifford algebra $Cl_{1,1}$ is a 4-dimensional algebra with basis:

$$\{1, \vec{e}_1, \vec{e}_2, \vec{e}_1\vec{e}_2\}$$

These basis elements are algebraically isomorphic to the matrices:

$$\left\{ 1 \cong \begin{bmatrix} 1 & 0 \\ 0 & 1 \end{bmatrix}, \vec{e}_1 \cong \begin{bmatrix} 0 & -1 \\ 1 & 0 \end{bmatrix}, \vec{e}_2 \cong \begin{bmatrix} 1 & 0 \\ 0 & -1 \end{bmatrix}, \vec{e}_1\vec{e}_2 \cong \begin{bmatrix} 0 & 1 \\ 1 & 0 \end{bmatrix} \right\} \quad (3.49)$$

We have:

$$\mathbb{OP}_s = \begin{bmatrix} a & -b \\ b & -a \end{bmatrix} \cong a\vec{e}_1 + b\vec{e}_2$$

$$\mathbb{S} = \begin{bmatrix} a & b \\ b & a \end{bmatrix} \cong a + b\vec{e}_1\vec{e}_2 \quad (3.50)$$

Complex Numbers The Higher Dimensional Forms – 2nd Edition

We will soon combine these intriguing odd parity matrices with complex numbers and hyperbolic complex numbers to make quaternions and dihedralions, but, before that, a small digression.

A Brief (and Conventional) Introduction to Quaternions - $\mathbb{H}$

The quaternions were discovered by the Irish mathematician William Hamilton. They were seen as an extension of complex numbers from two dimensions into four dimensions.

Quaternions are of the form: $q = a + \hat{i}b + \hat{j}c + \hat{k}d$. They are multiplicatively non-commutative, but they have all the other attributes of an algebraic field. Thus, they form a division algebra. We have $\hat{i}^2 = \hat{j}^2 = \hat{k}^2 = -1$ and the multiplicative relations $\hat{i}\hat{j} = \hat{k}, \hat{j}\hat{i} = -\hat{k}, \hat{j}\hat{k} = \hat{i}, \hat{k}\hat{j} = -\hat{i}, \hat{k}\hat{i} = \hat{j}, \hat{i}\hat{k} = -\hat{j}$. Addition and multiplication are what you would expect (taking account of the above multiplicative relations):

Addition:
$$\left(a + \hat{i}b + \hat{j}c + \hat{k}d\right) + \left(e + \hat{i}f + \hat{j}g + \hat{k}h\right) =$$
$$(a+e) + \hat{i}(b+f) + \hat{j}(c+g) + \hat{k}(d+h) \quad (3.51)$$

Multiplication:
$$\left(a + \hat{i}b + \hat{j}c + \hat{k}d\right)\left(e + \hat{i}f + \hat{j}g + \hat{k}h\right)$$
$$= a\left(e + \hat{i}f + \hat{j}g + \hat{k}h\right) + \hat{i}b\left(e + \hat{i}f + \hat{j}g + \hat{k}h\right) +$$
$$\hat{j}c\left(e + \hat{i}f + \hat{j}g + \hat{k}h\right) + \hat{k}d\left(e + \hat{i}f + \hat{j}g + \hat{k}h\right)$$
$$= ae + \hat{i}af + \hat{j}ag + \hat{k}ah + \hat{i}be + \hat{i}^2 bf + \hat{i}\hat{j}bg + \hat{i}\hat{k}bh +$$
$$\hat{j}ce + \hat{j}\hat{i}cf + \hat{j}^2 cg + \hat{j}\hat{k}ch + \hat{k}de + \hat{k}\hat{i}df + \hat{k}\hat{j}dg + \hat{k}^2 dh \quad (3.52)$$
$$= ae - bf - cg - dh + \hat{i}(af + be + ch - dg) +$$
$$\hat{j}(ag - bh + ce + df) + \hat{k}(ah + bg - cf + de)$$

Phew! The (conventional) conjugate of a quaternion $q = a + \hat{i}b + \hat{j}c + \hat{k}d$ is $q^* = a - \hat{i}b - \hat{j}c - \hat{k}d$, and division is done with the conjugate as with complex numbers:

Section 3

$$\frac{\left(a+\hat{i}b+\hat{j}c+\hat{k}d\right)}{\left(e+\hat{i}f+\hat{j}g+\hat{k}h\right)} = \frac{\left(a+\hat{i}b+\hat{j}c+\hat{k}d\right)\left(e-\hat{i}f-\hat{j}g-\hat{k}h\right)}{\left(e+\hat{i}f+\hat{j}g+\hat{k}h\right)\left(e-\hat{i}f-\hat{j}g-\hat{k}h\right)}$$

$$= \frac{\left(a+\hat{i}b+\hat{j}c+\hat{k}d\right)\left(e-\hat{i}f-\hat{j}h-\hat{k}h\right)}{e^2+f^2+g^2+h^2} \quad (3.53)$$

The (conventional) norm of quaternions is $Norm = a^2+b^2+c^2+d^2$.

In fact, dear old Bill found only a quarter of the story. We now relate the story in full. The odd parity matrices occur as 2×2 blocks within quaternion-like matrices. There are four types of quaternion-like algebras[53]. We have:

Matrix Form 1:

$$\begin{bmatrix} \mathbb{C}1 & \mathbb{OP}_c 1 \\ \mathbb{OP}_c 1 & \mathbb{C}1 \end{bmatrix} \triangleq \begin{bmatrix} a & b & c & d \\ -b & a & d & -c \\ c & d & a & b \\ d & -c & -b & a \end{bmatrix} \quad (3.54)$$

This matrix form is multiplicatively closed but not multiplicatively commutative. It is inversely closed with: $\det = \left(a^2+b^2-c^2-d^2\right)^2$ which is positive definite for all values of $\{a,b,c,d\}$. This is the algebra of a complex number and an odd parity number with the product:

$$\left(Ca, \mathbb{OP}_c b\right)\left(Cc, \mathbb{OP}_c d\right) = \left(\left(CaCc + \mathbb{OP}_c b \mathbb{OP}_c d\right), \left(Ca\mathbb{OP}_c d + Cc\mathbb{OP}_c b\right)\right) \quad (3.55)$$

We take the norm to be the fourth root of the determinant. In non-matrix notation, the number is: $a+\hat{i}b+\hat{j}c+\hat{k}d$. The conjugate is taken from the adjoint matrix, which is: $A^{Conjugate} = A^{-1}Det(A)$, which gives:

---

[53] 2nd edition note: When this was written, these odd parity matrices were not properly understood. There are two quaternion algebras and six A_3 algebras. See: Dennis Morris The Physics of Empty Space. It is now known that the six A_3 algebras contain our 4-dimensional space-time and both classical electromagnetism and General Relativity. See: Dennis Morris Upon General Relativity.

Complex Numbers The Higher Dimensional Forms – 2nd Edition

$$a(a^2+b^2-c^2-d^2)+\hat{ib}(-a^2-b^2+c^2+d^2)$$
$$+\hat{jc}(-a^2-b^2+c^2+d^2)+\hat{kd}(-a^2-b^2+c^2+d^2) \quad (3.56)$$

with relations:

$$\left\{\begin{array}{l}\hat{i}^2=-1,\ \hat{j}^2=+1,\ \hat{k}^2=+1,\ \hat{i}\hat{j}=-\hat{j}\hat{i}=\pm\hat{k},\\ \hat{i}\hat{k}=-\hat{k}\hat{i}=\pm\hat{j},\ \hat{j}\hat{k}=-\hat{k}\hat{j}=\pm\hat{i}\end{array}\right\} \quad (3.57)$$

Note: We could have taken the conjugate to be: $A^{Conjugate} = A^{-1}\sqrt{Det(A)}$, which is $a-\hat{ib}-\hat{jc}-\hat{kd}$. To match, we would then have taken the norm to be $Norm^2 = Det(A)$. Leading, via $A^{Conjugate}A = Norm^2$, to the same relations, (3.57).

Matrix Form 2:

$$\begin{bmatrix}\mathbb{S}1 & \mathbb{OP}_s 1 \\ \mathbb{OP}_s 1 & \mathbb{S}1\end{bmatrix} \triangleq \begin{bmatrix}a & b & c & -d \\ b & a & d & -c \\ c & -d & a & b \\ d & -c & b & a\end{bmatrix} \quad (3.58)$$

This matrix form is multiplicatively closed but not multiplicatively commutative. It is inversely closed with: $det = (a^2-b^2-c^2+d^2)^2$ which is positive definite for all values of $\{a,b,c,d\}$. This is the algebra of a hyperbolic complex number and an odd parity number with the product:

$$(\mathbb{S}a,\mathbb{S}b)(\mathbb{S}c,\mathbb{S}d) = ((\mathbb{S}a\mathbb{S}c+\mathbb{OP}_s b\mathbb{OP}_s d),(\mathbb{S}a\mathbb{OP}_s d+\mathbb{OP}_s b\mathbb{S}c)) \quad (3.59)$$

We take the norm to be the fourth root of the determinant. In non-matrix notation, the number is: $a+\hat{rb}+\hat{sc}+\hat{td}$. The conjugate is taken from the adjoint matrix:

$$a(a^2-b^2-c^2+d^2)+\hat{rb}(-a^2+b^2+c^2-d^2)$$
$$+\hat{sc}(-a^2+b^2+c^2-d^2)+\hat{td}(-a^2+b^2+c^2-d^2) \quad (3.60)$$

with relations:

51

$$\left\{ \begin{array}{c} \hat{r}^2 = +1,\ \hat{s}^2 = +1,\ \hat{t}^2 = -1,\ \hat{r}\hat{s} = -\hat{s}\hat{r} = \pm\hat{t}, \\ \hat{r}\hat{t} = -\hat{t}\hat{r} = \pm\hat{s},\ \ \hat{s}\hat{t} = -\hat{t}\hat{s} = \pm\hat{r} \end{array} \right\} \tag{3.61}$$

Note:

$$A^{Conjugate} = A^{-1}\sqrt{Det(A)} = a - \hat{i}b - \hat{j}c - \hat{k}d \tag{3.62}$$

and: $Norm^2 = Det(A)$. Lead to the same relations, (3.61).

<u>Matrix form 3:</u> - The quaternion algebra of William Hamilton.

$$\begin{bmatrix} \mathbb{C}1 & \mathbb{OP}_{\mathbb{C}}1 \\ -\mathbb{OP}_{\mathbb{C}}1 & \mathbb{C}1 \end{bmatrix} \triangleq \begin{bmatrix} a & b & c & d \\ -b & a & d & -c \\ -c & -d & a & b \\ -d & c & -b & a \end{bmatrix} \tag{3.63}$$

This matrix form is multiplicatively closed but not multiplicatively commutative. It is inversely closed with: $\det = \left(a^2 + b^2 + c^2 + d^2\right)^2$ which is positive definite for all values of $\{a,b,c,d\}$. This is the algebra of a complex number and an odd parity number with the product:

$$\left(\mathbb{C}a, \mathbb{OP}_{\mathbb{C}}b\right)\left(\mathbb{C}c, \mathbb{OP}_{\mathbb{C}}d\right) = \left(\left(\mathbb{C}a\mathbb{C}c - \mathbb{OP}_{\mathbb{C}}b\mathbb{OP}_{\mathbb{C}}d\right), \left(-\mathbb{C}a\mathbb{OP}_{\mathbb{C}}d - \mathbb{C}c\mathbb{OP}_{\mathbb{C}}b\right)\right)$$

(3.64)

We take the norm to be the fourth root of the determinant. In non-matrix notation, the number is: $a + \hat{i}b + \hat{j}c + \hat{k}d$. The conjugate is taken from the adjoint matrix:

$$\begin{array}{c} a\left(a^2 + b^2 + c^2 + d^2\right) + \hat{i}b\left(-a^2 - b^2 - c^2 - d^2\right) \\ + \hat{j}c\left(-a^2 - b^2 - c^2 - d^2\right) + \hat{k}d\left(-a^2 - b^2 - c^2 - d^2\right) \end{array} \tag{3.65}$$

with relations:

$$\left\{ \begin{array}{c} \hat{i}^2 = -1,\ \hat{j}^2 = -1,\ \hat{k}^2 = -1,\ \hat{i}\hat{j} = -\hat{j}\hat{i} = \pm\hat{k}, \\ \hat{i}\hat{k} = -\hat{k}\hat{i} = \pm\hat{j},\ \ \hat{j}\hat{k} = -\hat{k}\hat{j} = \pm\hat{i} \end{array} \right\} \tag{3.66}$$

Note: $A^{Conjugate} = A^{-1}\sqrt{Det(A)} = a - \hat{i}b - \hat{j}c - \hat{k}d$, and: $Norm^2 = Det(A)$. Lead to the same relations.

Matrix Form 4:

$$\begin{bmatrix} \mathbb{S}1 & \mathbb{OP}_s 1 \\ -\mathbb{OP}_s 1 & \mathbb{S}1 \end{bmatrix} \triangleq \begin{bmatrix} a & b & c & -d \\ b & a & d & -c \\ -c & d & a & b \\ -d & c & b & a \end{bmatrix} \quad (3.67)$$

This matrix form is multiplicatively closed but not multiplicatively commutative. It is inversely closed with: $\det = (a^2 - b^2 + c^2 - d^2)^2$ which is positive definite for all values of $\{a,b,c,d\}$. This is the algebra of a hyperbolic complex number and an odd parity number with the product:

$$(\mathbb{S}a, \mathbb{S}b)(\mathbb{S}c, \mathbb{S}d) = ((\mathbb{S}a\mathbb{S}c - \mathbb{OP}_s b\mathbb{OP}_s d), (-\mathbb{S}a\mathbb{OP}_s d - \mathbb{OP}_s b\mathbb{S}c)) \quad (3.68)$$

We take the norm to be the fourth root of the determinant. In non-matrix notation, the number is: $a + \hat{r}b + \hat{s}c + \hat{t}d$. The conjugate is taken from the adjoint matrix:

$$\begin{aligned} &a(a^2 - b^2 + c^2 - d^2) + \hat{r}b(-a^2 + b^2 - c^2 + d^2) \\ &+ \hat{s}c(-a^2 + b^2 - c^2 + d^2) + \hat{t}d(-a^2 + b^2 - c^2 + d^2) \end{aligned} \quad (3.69)$$

with relations:

$$\left\{ \begin{array}{l} \hat{r}^2 = +1,\ \hat{s}^2 = -1,\ \hat{t}^2 = +1,\ \hat{r}\hat{s} = -\hat{s}\hat{r} = \pm\hat{t}, \\ \hat{r}\hat{t} = -\hat{t}\hat{r} = \pm\hat{s},\ \hat{s}\hat{t} = -\hat{t}\hat{s} = \pm\hat{r}, \end{array} \right\} \quad (3.70)$$

Note: $A^{Conjugate} = A^{-1}\sqrt{Det(A)} = a - \hat{i}b - \hat{j}c - \hat{k}d$, and: $Norm^2 = Det(A)$. Lead to the same relations, (3.70).

Connections to the Clifford Algebras

Matrix Form 1 is the hyperbolic complex numbers with variables that are sums of parts of the $Cl_{2,0}$ Clifford algebra:

$$\mathbb{S}_{Cl_2}(a + b\vec{e}_1\vec{e}_2, c\vec{e}_1 + d\vec{e}_2)$$

Matrix Form 2 is the same but with elements from the Clifford algebra $Cl_{1,1}$

Section 3

$$\mathbb{S}_{Cl_{1,1}}\left(a+b\overrightarrow{e_1e_2}, c\overrightarrow{e_1}+d\overrightarrow{e_2}\right)$$

Matrix Form 3 is the Euclidean complex numbers with variables that are sums of parts of the $Cl_{2,0}$ Clifford algebra:

$$\mathbb{C}_{Cl_2}\left(a+b\overrightarrow{e_1e_2}, c\overrightarrow{e_1}+d\overrightarrow{e_2}\right)$$

Matrix Form 4 is the same but with elements from the Clifford algebra $Cl_{1,1}$

$$\mathbb{C}_{Cl_{1,1}}\left(a+b\overrightarrow{e_1e_2}, c\overrightarrow{e_1}+d\overrightarrow{e_2}\right)$$

Groups associated with these matrix forms:

The unit sub-matrices[54], together with their additive inverses, of the matrix form 3 (William Hamilton's quaternions) form the quaternion group, Q. The unit sub-matrices, with their additive inverses, of each of the other three matrix forms form the dihedral group, D_4. In this sense, there is only one quaternion algebra, and the other three algebras in this chapter are dihedral algebras. We appel the term di-hedralion to them[55].

The quaternion group has three C_4 subgroups and one C_2 subgroup. The C_2 subgroup is generated by the negative of the identity, of course. The three C_4 subgroups are generated by $\{B, C, D\}$. The dihedral group, D_4, has three C_2 subgroups and one C_4 subgroup. In each of these three dihedral algebras, the element that generates the single C_4 subgroup is different. In matrix form 1, the C_4 subgroup is generated by B. In matrix form 2, the C_4 subgroup is generated by D. In matrix form 4, the C_4 subgroup is generated by C. The three di-hedralion algebras are algebraically isomorphic. It is the sub-group structure that determines that there are three 4-dimensional dihedral algebras and only one 4-dimensional quaternion algebra.

---

[54] Take an algebraic matrix form, set one variable to unity, set the other variables to zero; you now have a unit sub-matrix.
[55] 2nd edition note: Ignore this reference to dihedralions. See: Dennis Morris The Physics of Empty Space.

Hyperbolic Complex Numbers are not algebraically closed:

Any polynomial with real or Euclidean complex numbers as coefficients can be factored into linear factors over the Euclidean complex numbers. This is called the algebraic closure of the Euclidean complex numbers.

Since $\hat{i} = \sqrt{-1}$ is not an element of the hyperbolic complex numbers, we cannot split:

$$\begin{bmatrix} x^2 & 0 \\ 0 & x^2 \end{bmatrix} + \begin{bmatrix} 1 & 0 \\ 0 & 1 \end{bmatrix} = \begin{bmatrix} 0 & 0 \\ 0 & 0 \end{bmatrix} \quad (3.71)$$

Into linear factors. The hyperbolic complex numbers are not algebraically closed. Of course, the real numbers are not algebraically closed either.

Vectors within Hyperbolic Space:

There is a one-to-one correspondence between the algebra of the hyperbolic complex numbers and 2-dimensional vectors.

$$\begin{bmatrix} a & b \\ b & a \end{bmatrix} \leftrightarrow \begin{bmatrix} a \\ b \end{bmatrix} \quad (3.72)$$

The angle operation is:

$$\begin{bmatrix} a & b \\ b & a \end{bmatrix} \odot \begin{bmatrix} c & d \\ d & c \end{bmatrix} = \begin{bmatrix} a & b \\ b & a \end{bmatrix} \begin{bmatrix} c & -d \\ -d & c \end{bmatrix}$$
$$= \begin{bmatrix} ac - bd & -(ad - bc) \\ -(ad - bc) & ac - bd \end{bmatrix} \quad (3.73)$$

Leading to:

$$\cosh \chi = \frac{ac - bd}{\sqrt{a^2 - b^2}\sqrt{c^2 - d^2}}$$
$$\sinh \chi = \frac{ad - bc}{\sqrt{a^2 - b^2}\sqrt{c^2 - d^2}} \quad (3.74)$$

---

[56] 2nd edition note: This was a huge insight seen through the wrong spectacles. The Clifford Algebras are the non-commutative algebras of the commutative $C_2 \times C_2 \times ...$ groups. See: Dennis Morris The Naked Spinor.

Section 3

The first of these is the established inner product of Minkowski space-time. Since $\cosh \chi$ is never zero, vectors in this space are never orthogonal.

The second of the above expressions is the magnitude of the wedge product in Minkowski space-time. Other than the normalising terms, the magnitude of the wedge product of Minkowski space-time is the same as the magnitude of the wedge product of 2-dimensional Euclidean space.

Differentiation of fields - hyperbolic complex numbers:

The differential of a scalar field over 2-dimensional hyperbolic space-time is:

$$d\begin{bmatrix} f(a,b) & 0 \\ 0 & f(a,b) \end{bmatrix} = \begin{bmatrix} \dfrac{\partial f}{\partial a} & \dfrac{\partial f}{\partial b} \\ \dfrac{\partial f}{\partial b} & \dfrac{\partial f}{\partial a} \end{bmatrix} \quad (3.75)$$

This is the gradient in 2-dimensional space-time.
The differential of a vector field is:

$$d\begin{bmatrix} F_a(a,b) & F_b(a,b) \\ F_b(a,b) & F_a(a,b) \end{bmatrix} = \begin{bmatrix} \dfrac{\partial F_a}{\partial a} + \dfrac{\partial F_b}{\partial b} & \dfrac{\partial F_b}{\partial a} + \dfrac{\partial F_a}{\partial b} \\ \dfrac{\partial F_b}{\partial a} + \dfrac{\partial F_a}{\partial b} & \dfrac{\partial F_a}{\partial a} + \dfrac{\partial F_b}{\partial b} \end{bmatrix}$$

$$= \begin{bmatrix} div(F) & curl(F) \\ curl(F) & div(F) \end{bmatrix}_{Space-time} \quad (3.76)$$

We see that each type of space has its own type of gradient, divergence, and curl. We see that whereas the curl of the Euclidean complex numbers was anti-symmetric in the matrix, the curl of space-time is symmetric in the matrix. Of course, the electromagnetic tensor is an anti-symmetric tensor and the gravitational tensor is a symmetric tensor[57].

2nd Edition note: We now know that these odd parity matrices are 2×2 representations of the 4-dimensional A_3 algebras from the $C_2 \times C_2$ group.

---

[57] 2nd Edition note: The electromagnetic tensor and energy-momentum tensor emerge from the 4-dimensional spaces that derive from the finite group $C_2 \times C_2$ - see "The Physics of Empty Space" by Dennis Morris. See: Upon General Relativity by Dennis Morris

Complex Numbers The Higher Dimensional Forms – 2nd Edition

Section 4

Introduction to Section 4:

Within this section we do a bit of tidying up. We meet 1-dimensional space from a different viewpoint, and we generalise 2-dimensional space.

1-dimensional Space:

The algebra of the real numbers is associated with 1-dimensional space. The real numbers have matrix form that is a 1×1 matrix. We call the 1-dimensional space the 'real number line'. There is only one 1-dimensional algebra – the real numbers, and so there is only one 1-dimensional space. This space is homogeneous in that the real numbers are the same distance apart at every position within this space. Isotropy is a meaningless concept in 1-dimensional space. There is no concept of angle in 1-dimensional space.

The polar form of the real numbers is:
$$\exp([a]) = \exp([a]+[0])$$
$$= \exp([a])\exp([0]) \quad (4.1)$$
$$= [e^a][1]$$

$[1]$ is the 'rotation matrix' of 1-dimensional space. We jump the gun a little, bend the terminology a little, and declare that this rotation matrix contains the 1-dimensional trigonometric function, 1. This function takes zero arguments.

Mathematicians are wont to construct higher dimensional spaces by fixing together a number of 1-dimensional spaces. Such spaces are not a pretty sight. However, we will demonstrate the procedure.
a) Assume the existence of n axes that are n 1-dimensional spaces.
b) Fix them together at the point zero, which will hereafter be called the origin.
c) Within these 1-dimensional spaces, there is no concept of n-dimensional distance. Such a concept is needed, and so the mathematician invents it. The mathematician won't accept any old n-dimensional distance function; it is required to satisfy various axioms (usually the metric space axioms), but, none-the-less, the function comes from within the mathematician's head not from within the algebra.

c) There is no concept of angle within 1-dimensional space. Such a concept is needed, and so the mathematician invents it. It is called the inner product, and it is essentially a projection from one line within the space on to another line in the space. If this projection is zero, then the lines are orthogonal to each other. Often, but not necessarily, the inner product is partly derived from the invented distance function. The inner product cannot be wholly derived from the distance function because its construction requires a trigonometric function of some kind that accepts an argument – an angle. From where does the mathematician acquire the required trigonometric function? It is kidnapped from a 2-dimensional space. Usually, it is the cos trigonometric function that is kidnapped, but the cosh function is kidnapped sometimes.

And there we have it:

Conventional *n*-dimensional space:
$\quad n$ 1-dimensional spaces
$\quad$ An invented distance function
$\quad$ An invented inner-product
$\quad$ A kidnapped trigonometric function

2nd edition note: We now know that trigonometric functions occur in only complex number types of space (spinor spaces). The higher dimensional trigonometric functions are destroyed by super-imposition of the higher dimensional algebras to form our classical universe, but the 2-dimensional trigonometric functions, together with the whole of the two 2-dimensional algebras survive super-imposition because they occur in only a single basis each. This is why we see 2-dimensional rotations in our 4-dimensional space-time but do not see 4-dimensional rotations in our 4-dimensional space-time. There are 4-dimensional rotations in quantum physics (electron spin), but we cannot perceive them as 4-dimensional rotations from within our classical 4-dimensional space-time. See "The Physics of Empty Space" by Dennis Morris.

2nd Edition note: When a mathematician chooses a distance function, she automatically and surreptitiously chooses the type of rotation which will preserve that distance function or parts of that distance function. Thus, she automatically chooses the type of angles in the space and the numbers of them.

Section 4

The General Rotation Matrix – 2-dimensions:

The general form of the 2-dimensional algebraic matrix form is:

$$\begin{bmatrix} a & b \\ jb & a \end{bmatrix} : j \neq 0 \qquad (4.2)$$

We have:

$$\exp\left(\begin{bmatrix} a & b \\ jb & a \end{bmatrix}\right) = \begin{bmatrix} e^a & 0 \\ 0 & e^a \end{bmatrix} \begin{bmatrix} \cosh(\sqrt{jb}) & \frac{1}{\sqrt{j}}\sinh(\sqrt{jb}) \\ j\frac{1}{\sqrt{j}}\sinh(\sqrt{jb}) & \cosh(\sqrt{jb}) \end{bmatrix} \qquad (4.3)$$

The trigonometric matrix is the rotation matrix of the 2-dimensional natural space. When $j = 1$, we have the two matrices corresponding to the two roots of unity:

$$\left\{ \begin{array}{l} \begin{bmatrix} \cosh(b) & \sinh(b) \\ \sinh(b) & \cosh(b) \end{bmatrix} \\ \begin{bmatrix} \cosh(-b) & -\sinh(-b) \\ -\sinh(-b) & \cosh(-b) \end{bmatrix} = \begin{bmatrix} \cosh(b) & \sinh(b) \\ \sinh(b) & \cosh(b) \end{bmatrix} \end{array} \right. \qquad (4.4)$$

When $j = -1$:

$$\left\{ \begin{array}{l} \begin{bmatrix} \cosh(\hat{\imath}b) & -\hat{\imath}\sinh(\hat{\imath}b) \\ \hat{\imath}\sinh(\hat{\imath}b) & \cosh(\hat{\imath}b) \end{bmatrix} = \begin{bmatrix} \cos(b) & \sin(b) \\ -\sin(b) & \cos(b) \end{bmatrix} \\ \begin{bmatrix} \cosh(-\hat{\imath}b) & \hat{\imath}\sinh(-\hat{\imath}b) \\ -\hat{\imath}\sinh(-\hat{\imath}b) & \cosh(-\hat{\imath}b) \end{bmatrix} = \begin{bmatrix} \cos(-b) & -\sin(-b) \\ \sin(-b) & \cos(-b) \end{bmatrix} = \begin{bmatrix} \cos(b) & \sin(b) \\ -\sin(b) & \cos(b) \end{bmatrix} \end{array} \right.$$

(4.5)

Where we have used the well known relations[58]:

$$\cosh(\hat{\imath}b) = \cos(b)$$
$$\sinh(\hat{\imath}b) = \hat{\imath}\sin(b) \qquad (4.6)$$

The general 2-dimensional conjugate is:

---

[58] Alternatively, we already know the cases for plus and minus unity, and, from these, we could deduce the well-known relations.

$$\begin{bmatrix} a & -b \\ -jb & a \end{bmatrix} \quad (4.7)$$

Thus the general angle product is:

$$\begin{bmatrix} a & b \\ jb & a \end{bmatrix} \odot \begin{bmatrix} c & -d \\ -jd & c \end{bmatrix} = \begin{bmatrix} ac - jbd & -(ad - bc) \\ -j(ad - bc) & ac - jbd \end{bmatrix} \quad (4.8)$$

Leading to the 2-dimensional general dot product:

$$\cosh\left(\sqrt{j}\theta\right) = \frac{ac - jbd}{\sqrt{a^2 - jb^2}\sqrt{c^2 - jd^2}} \quad (4.9)$$

and the general 2-dimensional wedge product:

$$\frac{1}{\sqrt{j}}\sinh\left(\sqrt{j}\theta\right) = \frac{ad - bc}{\sqrt{a^2 - jb^2}\sqrt{c^2 - jd^2}} \quad (4.10)$$

The general distance function is:

$$r = \sqrt[2]{a^2 - jb^2} \quad (4.11)$$

and so space is squashed by $\sqrt{j}$ in the b direction[59].

---

[59] 2nd edition note: We now know that the 'space squeezing' parameters like j are physical constants. In the hyperbolic complex numbers, the 'space squeezing' parameter is the limiting velocity of the universe, In the Euclidean complex numbers the 'space squeezing' parameter is Planck's constant.

Section 5

Introduction to Section 5:

In previous sections we have met two 2-dimensional algebraic matrix forms, the Euclidean complex numbers and the hyperbolic complex numbers, and the one 1-dimensional algebraic matrix form that is the real numbers. Clearly, we wonder if these are the only algebraic matrix forms or are there other 2-dimensional ones or others of higher dimension.

In this section, we unveil the structures of the matrix forms of the other natural algebraic fields. We do this be deducing from the algebraic field axioms[60] the properties of algebraic matrix forms in general.

The Algebraic Matrix Forms from the Field Axioms:

The axioms of an algebraic field are given in Appendix 2. In this work, we are considering only sets of matrices with matrix addition and matrix multiplication. Since any set of matrices with matrix addition and matrix multiplication automatically satisfies many of the axioms of an algebraic field, we do not need to consider every single one of the field axioms. We have to consider only:

The Axioms of an Algebraic Matrix Field:

F3: Inclusion of the additive identity, $\begin{bmatrix} 0 & 0 \\ 0 & 0 \end{bmatrix}$[61].
F4: Inclusion of additive inverses
F5: Additive Closure
F6: Multiplicative Closure
F7: Inclusion of multiplicative inverses
F8: Inclusion of the multiplicative identity, $\begin{bmatrix} 1 & 0 \\ 0 & 1 \end{bmatrix}$.
F11: Multiplication is commutative
F14: There are no non-singular matrices in the set

---

[60] See Appendix 2

[61] We demonstrate with 2×2 matrices, but the same applies to larger matrices

The Characterization of Matrix Forms:

A matrix form may be such that all its elements are independent of each other or some of its elements are dependent upon other of its elements. We characterize such a matrix form by specifying the relationships between the elements or by writing the matrix; for example:

$$\{a_{11} \in \mathbb{R}, a_{12} \in \mathbb{R}, a_{21} = -a_{12}, a_{22} = 3\} \equiv \begin{bmatrix} a & b \\ -b & 3 \end{bmatrix} \qquad (5.1)$$

Theorem 1
Algebraic matrix forms must be square matrices.
Proof:
The requirement for multiplicative commutativity.

The required inclusion of the multiplicative identity together with the requirement for additive closure allows us to build scalars.

$$\begin{bmatrix} 1 & 0 \\ 0 & 1 \end{bmatrix} + \begin{bmatrix} 1 & 0 \\ 0 & 1 \end{bmatrix} = \begin{bmatrix} 2 & 0 \\ 0 & 2 \end{bmatrix} \qquad (5.2)$$

Multiplicative closure[62] ensures that multiplication by a scalar is inclusive.

Theorem 2
The elements of algebraic matrix forms cannot be (or contain) non-zero constants.
Proof:
The requirement for multiplicative closure after multiplication by a scalar.

Theorem 3
The elements of algebraic matrix forms must be either independent of other elements or linear functions of the other elements.
Proof:
The requirement for multiplicative closure after multiplication by a scalar.

We demonstrate what is stated in Theorem 3 is necessary to maintain the form of the matrix.

$$\begin{bmatrix} 2 & 0 \\ 0 & 2 \end{bmatrix} \begin{bmatrix} a & 0 \\ 0 & a^2 \end{bmatrix} = \begin{bmatrix} 2a & 0 \\ 0 & 2a^2 \end{bmatrix} \qquad (5.3)$$

---

[62] Additive/Multiplicative closure means that the product matrix must be of the same form as the matrices from which it is made.

Section 5

Theorem 3 means that all the elements of an algebraic matrix form are either independent, $a_{ij} \in \mathbb{R}$, of other elements or are of the form:

$$a_{st} = ja_{zy} + ka_{xw} + \ldots : \{j, k, \ldots\} \in \mathbb{R} \qquad (5.4)$$

Theorem 4
The elements on the leading diagonal[63] of an algebraic matrix form must be independent of all elements that are not on the leading diagonal.
Proof:
The requirement for inclusion of the multiplicative identity and Theorem 3.

Theorem 5
The elements on the leading diagonal of an algebraic matrix form are all dependent upon each other. (From Theorem 3, the dependence is linear.)
Proof:
We require guaranteed non-singularity for every non-zero matrix. In particular we require non-singularity when all variables within a matrix are zero except one. If the leading diagonal has different variables upon it, the requirement for guaranteed non-singularity cannot be met. (Remember Theorem 4).

Although theorem 5 is valid, we could avoid non-singularity by imposing restrictions that disallow any leading diagonal element from being zero unless all such elements are zero.[64] If the reader is not satisfied with Theorem 5, we can put things differently:

A Different Theorem 5
The elements on the leading diagonal of an algebraic matrix form are all equal to each other.
Reason:
We require this in order that we can include the real numbers as a sub-algebra. Without such inclusion, the geometric spaces within the algebras will not have a 1-dimensional sub-space.

---

[63] The leading diagonal of a square matrix is the diagonal that runs from the top left-hand corner to the bottom right-hand corner.
[64] This is exactly what we will do when we consider matrices of the form:

$$\begin{bmatrix} a & b & 0 \\ -b & a & 0 \\ 0 & 0 & c \end{bmatrix}$$

Theorem 6
The elements on the leading diagonal of an algebraic matrix form are all equal.
Proof:
Theorem 3 with Theorem 5 and the requirement to include the multiplicative identity.

Corollary:
Every algebraic matrix form contains, as a sub-algebra, an algebra that is isomorphic to the real numbers.
Proof:
Matrices that have all elements of the leading diagonal equal and all other element zero are algebraically isomorphic to the real numbers.

Theorem 7
The only dependency relation is of the form $a_{st} = ja_{xy} : j \in \mathbb{R}$
Proof:
The requirement of the inclusion of the multiplicative inverse and that the elements on the leading diagonal of the multiplicative inverse must be equal – Theorem 6 – and independent of the other elements – Theorem 4.

Theorem 7 over-reaches Theorem 3, which is now redundant.

Theorem 8
For an $n \times n$ matrix with n independent variables, each variable or a multiple of each variable in an algebraic matrix form must occur once and only once in each column and in each row of the matrix.
Proof:
A matrix form can be an algebra only if the matrix form is guaranteed non-singular. This guarantee needs to apply when all but one of the variables within the matrix form is zero. A matrix with a row or column of zeros is always singular.
Alternative Proof:
This is required in order to maintain equality of the elements on the leading diagonal under multiplication.

Theorem 8 deals with $n \times n$ matrices with n independent variables. $n \times n$ matrices with less than n independent variables can be deduced from the n independent variables matrices by simply tying two or more variables together. We have no interest in such matrices. Theorem 8 is very restrictive. It is sufficiently restrictive to serve our needs.

Section 5

We are now in a position to begin forming prospective algebraic matrix forms. We are guided by:
a) The leading diagonal elements are all equal – Theorem 6
b) A $n \times n$ algebraic matrix form has n independent variables, and these variables each occur once and only once in every row and in every column of the algebraic matrix form - Theorem 8.
c) The only dependency relation is of the form $a_{st} = ja_{xy} : j \in \mathbb{R}$ - Theorem 7

How to Construct Algebraic Matrix Forms
1) Choose the size of the $n \times n$ square matrix.
2) Fill the leading diagonal with a's.
$$\begin{bmatrix} a & \\ & a \end{bmatrix}$$
3) Find every possible matrix with $(n-1)$ other independent variables arranged so that each variable occurs once and only once in every column and in every row of the matrix.
$$\begin{bmatrix} a & b \\ b & a \end{bmatrix}$$
4) Other than the leading diagonal and the top row, multiply the other elements of the matrix by a (space squeezing) parameter
$$\begin{bmatrix} a & b \\ jb & a \end{bmatrix}$$
5) Multiply the matrix thus formed by itself and adjust the parameters to ensure the form of the matrix is maintained by multiplication.
$$\begin{bmatrix} a & b \\ jb & a \end{bmatrix}^2 = \begin{bmatrix} a^2 + jb^2 & 2ab \\ 2jab & a^2 + jb^2 \end{bmatrix}$$
(The case of the 3×3 matrices is below.)

Nomenclature[65]:   *Space Squeezing Parameters*
The parameters that occur in algebraic matrix forms that determine the dependency of one element upon another are called space squeezing parameters.

---

[65] 2nd edition note: Space-squeezing parameters are now referred to as scaling parameters.

Having so formed prospective algebraic matrix forms in accordance with the above, we will have to check the algebraic matrix field axioms to ensure that they do form an algebraic field.

F3 is automatically satisfied for matrices whose elements are unrestricted real variables that are either independent or are linear multiples of each other.

F4 is automatically satisfied for matrices whose elements are unrestricted real variables.

F5 is automatically satisfied for matrices whose elements are unrestricted real variables that are either independent or are linear multiples of each other.

F6 is automatically satisfied for square matrices whose elements are independent real variables that appear once and only once in each row and in each column[66].

F7 is automatically satisfied for square matrices whose elements are independent real variables that appear once and only once in each row and in each column provided F14 is also satisfied.

F8 is automatically satisfied for matrices whose leading diagonal elements are all equal real variables that are independent of all other elements

This leaves us with only axioms F11 & F14 to check.

F11: There are matrices of the form we have constructed that are not multiplicatively commutative; the quaternion matrix form is one example.

F14: There are matrices of the form we have constructed whose cartesian forms are not guaranteed non-singular.

Theorem 9[67]

Multiplicatively commutative matrices of the form we have constructed are guaranteed non-singular in their polar form.

Proof:

Since the matrices are multiplicatively commutative for all values of the variables, then they are multiplicatively commutative when all variables except one are zero. We can therefore split the exponentiated form of these matrices into exponentiated separate variable sub-matrices. One of these sub-matrices will be the sub-matrix with the leading diagonal variable; the other sub-matrices will have trace equal to zero. Thus the determinant of the polar form of the matrix will never be singular.

---

[66] 2nd edition note: This is an error. F6 is not automatically satisfied, but the proof stands.

[67] If this theorem sounds confusing now, don't worry. It will become crystal clear later in the work.

Section 5

We have seen theorem 9 in action when we took the polar form of the hyperbolic complex numbers. We now demonstrate theorem 9 in a 3-dimensional case:

$$\exp\left(\begin{bmatrix} a & b & c \\ c & a & b \\ b & c & a \end{bmatrix}\right) = \exp\left(\begin{bmatrix} a & 0 & 0 \\ 0 & a & 0 \\ 0 & 0 & a \end{bmatrix} + \begin{bmatrix} 0 & b & 0 \\ 0 & 0 & b \\ b & 0 & 0 \end{bmatrix} + \begin{bmatrix} 0 & 0 & c \\ c & 0 & 0 \\ 0 & c & 0 \end{bmatrix}\right)$$

$$= \exp\left(\begin{bmatrix} a & 0 & 0 \\ 0 & a & 0 \\ 0 & 0 & a \end{bmatrix}\right)\exp\left(\begin{bmatrix} 0 & b & 0 \\ 0 & 0 & b \\ b & 0 & 0 \end{bmatrix}\right)\exp\left(\begin{bmatrix} 0 & 0 & c \\ c & 0 & 0 \\ 0 & c & 0 \end{bmatrix}\right)$$

(5.5)

$$\det\left(\exp\left(\begin{bmatrix} a & b & c \\ c & a & b \\ b & c & a \end{bmatrix}\right)\right)$$

$$= \det\left(\begin{bmatrix} e^a & 0 & 0 \\ 0 & e^a & 0 \\ 0 & 0 & e^a \end{bmatrix}\right)\det\left(\exp\left(\begin{bmatrix} 0 & b & 0 \\ 0 & 0 & b \\ b & 0 & 0 \end{bmatrix}\right)\right)\det\left(\exp\left(\begin{bmatrix} 0 & 0 & c \\ c & 0 & 0 \\ 0 & c & 0 \end{bmatrix}\right)\right)$$

$$= e^{3a}.\exp\left(Trace\begin{bmatrix} 0 & b & 0 \\ 0 & 0 & b \\ b & 0 & 0 \end{bmatrix}\right)\exp\left(Trace\begin{bmatrix} 0 & 0 & c \\ c & 0 & 0 \\ 0 & c & 0 \end{bmatrix}\right)$$

$$= e^{3a}.1.1$$

(5.6)

Thus, if we use the polar forms (or only matrices equivalent to a polar form), we can ignore F14.[68] We need to consider only F11 - multiplicative commutativity. This is what we will do.

Important Point: In each case, the set of matrices that form the algebra is the set of matrices that are equal to the polar form matrices of the algebra. In general, this is not all of the set of the Cartesian matrices. In the case of the Euclidean complex numbers it is the whole Cartesian set. In the case of the hyperbolic complex numbers it is a restricted part of the Cartesian set. Thus, we avoid having to impose restrictions upon the values that the variables within the algebra can take.

Theorem 10

---

[68] Which is one huge headache out of the way!

Matrices constructed as detailed in 'How to Construct Algebraic Matrix Forms' above are algebraic fields if they are multiplicatively commutative.
Proof: Given above
An Aside: If such matrices are not multiplicatively commutative, we cannot form the polar form and thereby be rid of the requirement for non-singularity. However, such matrices will form a multiplicatively non-commutative division algebra with appropriate restrictions applied to their Cartesian form. The quaternions are atypical in that they require no restrictions.

The 1×1 Matrices:

The 1×1 matrices are the real numbers, and these are a field.

The 2×2 Matrices:

The only prospective form is:

$$\begin{bmatrix} a & b \\ jb & a \end{bmatrix} : j \in \mathbb{R} \qquad (5.7)$$

We need to impose the restriction $|a| > |\sqrt{jb}|$ to make this into an algebraic field and thus an algebraic matrix form.

Most often, we will be concerned only with the cases $j = \pm 1$ (and similarly for other such parameters). We call such algebras flat natural algebras.

2nd edition note: When the 1st edition of this book was written, it was anticipated that it would be possible to curve these spinor spaces (complex number spaces) by allowing the parameters like j to vary from point to point. It is now known that this cannot be done without destroying the algebraic structure, and so all spinor spaces are flat spaces. However, the $\mathbb{R}^n$ spaces that arise from superimposition of isomorphic spinor spaces can be curved because they are not algebraic structures. By now, the reader knows where to look for details.

Nomenclature: *Flat Algebras or Flat Algebraic Fields*[69]
An algebraic matrix form with spaces squeezing parameters equal to plus or minus unity.

When $j = -1$, we have:

---

[69] 2nd edition note: A rather embarrassing use of the word flat. All spinor algebra spaces are flat. Only emergent expectation spaces can be curved.

Section 5

$$\mathbb{C} \cong \begin{bmatrix} a & b \\ -b & a \end{bmatrix} \quad (5.8)$$

This is the complex numbers in matrix notation. The case $j=1$ gives:

$$\mathbb{S} \cong \begin{bmatrix} a & b \\ b & a \end{bmatrix} : |a| > |b| \quad (5.9)$$

This is the hyperbolic complex numbers.

We put all this into a theorem.

Theorem 11
The only forms of 2×2 matrices that are an algebraic field are[70]:

$$\left\{ \begin{bmatrix} a & b \\ jb & a \end{bmatrix} : j<0;\ \begin{bmatrix} a & b \\ 0 & a \end{bmatrix} : a \neq 0;\ \begin{bmatrix} a & b \\ jb & a \end{bmatrix} : |a|>|b|, j>0 \right\} \quad (5.10)$$

Proof: Given above

Corollary: With $j=\pm 1$, the only forms of 2×2 matrices that are a flat algebraic field are:

$$\left\{ \mathbb{C} \cong \begin{bmatrix} a & b \\ -b & a \end{bmatrix}, \mathbb{S} \cong \begin{bmatrix} a & b \\ b & a \end{bmatrix} : |a|>|b| \right\} \quad (5.11)$$

The 3×3 Matrices:
The only prospective algebraic matrix form is:

$$\begin{bmatrix} a & b & c \\ jc & a & kb \\ lb & mc & a \end{bmatrix} \quad (5.12)$$

We now use the requirement of equality of leading diagonal elements together with the requirement of multiplicative closure. We multiply this matrix by itself and equate the leading diagonal elements of the product[71].

---

[70] We count the real numbers as a 1×1 matrix.
[71] With larger matrices, it is necessary to consider the off-diagonal elements also.

$$\begin{bmatrix} a & b & c \\ jc & a & kb \\ lb & mc & a \end{bmatrix}^2 = \begin{bmatrix} a^2+jbc+lbc & \sim & \sim \\ \sim & a^2+jbc+kmbc & \sim \\ \sim & \sim & a^2+kmbc+lbc \end{bmatrix}$$
(5.13)

From this we get:

$$\begin{bmatrix} a & b & c \\ jc & a & kb \\ jb & \dfrac{j}{k}c & a \end{bmatrix} : j \in \mathbb{R}, k \in \mathbb{R} \neq 0 \quad (5.14)$$

Testing the other field axioms, this is (in Cartesian form with restrictions[72]) an algebraic field. Although (with restrictions) this is an algebraic field for the stated values of $\{j,k\}$, our only concern at present is with the flat algebras. These are given by the four permutations of $\{j,k\} = \pm 1$.

Theorem 12
The only forms of 3×3 matrices with three independent variables that are an algebraic field are (with appropriate restrictions):

$$C_3 = \begin{bmatrix} a & b & c \\ jc & a & kb \\ jb & \left(\dfrac{j}{k}\right)c & a \end{bmatrix} : j \in \mathbb{R}, k \in \mathbb{R} \neq 0 \quad (5.15)$$

Proof: Given above.

Corollary: With $\{j,k\} = \pm 1$, the only forms of 3×3 matrices (with three independent variables) that are an algebraic field are (with restrictions):

---

[72] We do not dwell upon the restrictions; they are of no importance because the polar forms do not need them. They simply reflect the difficulty of trying to make non-Euclidean spaces match the extent of Euclidean space.

Section 5

$$\left\{\begin{array}{c}C_3L^1H^2{}_{(j=1,k=-1)} = \begin{bmatrix} a & b & c \\ c & a & b \\ b & c & a \end{bmatrix}, \quad C_3L^1E^1H^1{}_{(j=-1,k=-1)} = \begin{bmatrix} a & b & c \\ -c & a & b \\ -b & -c & a \end{bmatrix} \\ \\ C_3L^1E^2{}_{(j=1,k=-1)} = \begin{bmatrix} a & b & c \\ c & a & -b \\ b & -c & a \end{bmatrix}, \quad C_3L^1E^1H^1{}_{(j=-1,k=-1)} = \begin{bmatrix} a & b & c \\ -c & a & -b \\ -b & c & a \end{bmatrix} \end{array}\right\} \quad (5.16)$$

The reader who is familiar with circulant matrices[73] and skew circulant matrices will see two familiar matrices here. If the reader has never heard of circulant matrices, do not worry; it will be explained later.

The names we have appelled to these algebras will seem strange. We provide a pre-emptive explanation below

---

The Origin of the Names Appelled to these Algebras

The name of an algebra has three components.
i) The name of the group that underlies the algebra. This will mean nothing to the reader yet – it is explained later. If there is more than one algebra derived from a particular group (because sub-groups are generated by different elements of the group), then these different forms are denoted by a decimal point and a number added to the name of the group. Eg: $C_{4.1}$

ii) The polar form of the algebra (if known). Eg: L^1H^3. The L is the (always present) length matrix. The H stands for a rotation matrix with H-type simple-trig functions within it. The simple-trig functions will be introduced later. The E stands for a rotation matrix with E-type simple-trig functions within it.

iii) The subscripted values of the space squeezing parameters that generate the flat form of the algebra. Eg: $_{(j=1,k=-1)}$

4×4 Matrices:

The prospective algebraic matrix forms (with four independent variables) are:

---

[73] See appendix 4

Complex Numbers The Higher Dimensional Forms – 2nd Edition

$$\left\{ \begin{bmatrix} a & b & c & d \\ jb & a & kd & lc \\ mc & nd & a & pb \\ qd & rc & sb & a \end{bmatrix}, \begin{bmatrix} a & b & c & d \\ jb & a & kd & lc \\ md & nc & a & pb \\ qc & rd & sb & a \end{bmatrix}, \right. \\ \left. \begin{bmatrix} a & b & c & d \\ jd & a & kb & lc \\ mc & nd & a & pb \\ qb & rc & sd & a \end{bmatrix}, \begin{bmatrix} a & b & c & d \\ jc & a & kd & lb \\ mb & nd & a & pc \\ qd & rc & sb & a \end{bmatrix} \right\} \quad (5.17)$$

In each case, we multiply this matrix by itself and equate the leading diagonal elements of the product (and a bit more).[74] This leads to:

$$\left\{ C_{41} = \begin{bmatrix} a & b & c & d \\ jd & a & kb & lc \\ \frac{jl}{k}c & \frac{j}{k}d & a & lb \\ jb & \frac{j}{k}c & \frac{j}{l}d & a \end{bmatrix} : \{k,l\} \neq 0, \; C_{42} = \begin{bmatrix} a & b & c & d \\ jb & a & kd & \frac{j}{k}c \\ ld & \frac{l}{k}c & a & \frac{j}{k}b \\ lc & \frac{kl}{j}d & kb & a \end{bmatrix} : \{j,k\} \neq 0, \right. \\ \left. C_{43} = \begin{bmatrix} a & b & c & d \\ jc & a & kd & lb \\ jb & \frac{j}{l}d & a & \frac{j}{k}c \\ \frac{jk}{l}d & \frac{j}{l}c & kb & a \end{bmatrix} : \{k,l\} \neq 0, C_2 \times C_2 = \begin{bmatrix} a & b & c & d \\ jb & a & \frac{j}{l}d & lc \\ kc & \frac{k}{l}d & a & lb \\ \frac{jk}{l^2}d & \frac{k}{l}c & \frac{j}{l}b & a \end{bmatrix} : \{l\} \neq 0 \right\}$$

(5.18)

We check the other field axioms and find that (with restrictions in the Cartesian form) all these matrix forms, with all permutations of $\{j,k,l\} = \{+1,-1\}$, are algebraic fields. The first three, $\{C_{4\cdot 1..3}\}$, are

---

[74] When calculating larger algebraic matrix forms, setting the leading diagonal elements equal is insufficient to complete the calculation and we need to look at the structure of the whole matrix. If we were brave, we could exponentiate the general form sub-matrices off the leading diagonal, take the determinants of those exponentiated matrices, and set parameters in a way that cancels the parameters from within those determinants.

Section 5

algebraically isomorphic to each other[75]. The fourth is a distinct algebra – it is two algebras fixed together in a groupy[76] sort of way.

We have now done enough to introduce the reader to algebraic matrix forms in their Cartesian form. In due course, we will discover that the polar forms of these algebras are more useful and more interesting. We emphasize that, in the polar form, we do not need to impose restrictions, and this is the reason for our lack of concern about the imposed restrictions.

The above given algebraic matrix forms are the only flat $\{1\times 1, 2\times 2, 3\times 3, 4\times 4\}$ matrix forms (with n independent variables) that satisfy the requirement to be an algebraic field. Thus, these are the only 'small' natural algebras. They include two types of 2-dimensional complex numbers, four types of 3-dimensional complex numbers, and, up to isomorphism, eight types of 4-dimensional complex numbers. They also include four algebras that are comprised of 2×2 blocks of 2-dimensional algebras and, with them, four other algebras.

Jumping the Gun:

The immediately preceding paragraph is not quite the true situation. It turns out that the n-dimensional algebraic matrix forms with one or more negative space squeezing parameters are folded forms of the $2n$-dimensional algebraic matrix form with all positive space squeezing parameters. Thus it is that there is only one 1-dimensional flat algebraic matrix form, only one 2-dimensional flat algebraic matrix form, only one 3-dimensional flat algebraic matrix form, etc.. However, it is too early in this work to introduce the concept of folded algebras. It will be introduced later.

Addendum 1 to Chapter 1

Block Algebras:

Because of the block multiplication nature of matrices, larger matrices that incorporate algebraic matrix forms as blocks upon the leading diagonal have many of the properties of an algebra. They are most often not a bona-fide algebraic matrix form because the leading diagonal elements are not kept equal; there are exceptions. We present some examples:

---

[75] Proof is given as an addendum to this chapter.
[76] Precise and clear prose is indicative of a writer who presumes the reader is of insufficient intelligence to understand sloppy prose. One can combine algebras in a non-groupy way – put them on the leading diagonal.

$$\begin{bmatrix} a & b & 0 & 0 \\ -b & a & 0 & 0 \\ 0 & 0 & a & b \\ 0 & 0 & b & a \end{bmatrix}^2 = \begin{bmatrix} a^2-b^2 & 2ab & 0 & 0 \\ -2ab & a^2-b^2 & 0 & 0 \\ 0 & 0 & a^2+b^2 & 2ab \\ 0 & 0 & 2ab & a^2+b^2 \end{bmatrix} \quad (5.19)$$

$$\begin{bmatrix} a & b & c & 0 \\ c & a & b & 0 \\ b & c & a & 0 \\ 0 & 0 & 0 & 1 \end{bmatrix}^2 = \begin{bmatrix} a^2+2bc & c^2+2ab & b^2+2ac & 0 \\ b^2+2ac & a^2+2bc & c^2+2ab & 0 \\ c^2+2ab & b^2+2ac & a^2+2bc & 0 \\ 0 & 0 & 0 & 1 \end{bmatrix} \quad (5.20)$$

Addendum 2 to Chapter 1

Proof of the Algebraic Isomorphism of the three 4-dimension C_4 type Algebras:

Note: In these proofs, we use the non-matrix form of notation.[77] The multiplication operations can be found by inspection of the product matrix of two matrices of a particular algebra.

Theorem:

$C_{4.1}$ is isomorphic as an algebra to $C_{4.2}$.

Proof:

$$\phi(C_{4.1} \mapsto C_{4.2}): \left(a+b\hat{r}+c\hat{s}+d\hat{t} \mapsto a+c\hat{r}+b\hat{s}+b\hat{t}\right)$$

Addition:

$$\phi\begin{pmatrix} \left(a+b\hat{r}+c\hat{s}+d\hat{t}\right)+\left(e+f\hat{r}+g\hat{s}+h\hat{t}\right) \\ =(a+e)+(b+f)\hat{r}+(c+g)\hat{s}+(d+h)\hat{t} \end{pmatrix}$$

$$\mapsto \begin{pmatrix} (a+e)+(c+g)\hat{r}+(b+f)\hat{s}+(d+h)\hat{t} \\ =\left(a+c\hat{r}+b\hat{s}+d\hat{t}\right)+\left(e+g\hat{r}+f\hat{s}+h\hat{t}\right) \end{pmatrix}$$

The Multiplication:
The multiplication operations within these two algebras are different.

---

[77] The non-matrix form of notation is yet to be dealt with. The reader might like to return to this addendum on a second reading.

Section 5

$$\phi \begin{pmatrix} \left(a+b\hat{r}+c\hat{s}+d\hat{t}\right) \times C_{4.1} \times \left(e+f\hat{r}+g\hat{s}+h\hat{t}\right) \\ = (ae+bh+cg+df)+(af+be+ch+dg)\hat{r} \\ +(ag+bf+ce+dh)\hat{s}+(ah+bg+cf+de)\hat{t} \end{pmatrix}$$

$$\mapsto \begin{pmatrix} (ae+bh+cg+df)+(ag+bf+ce+dh)\hat{r} \\ +(af+be+ch+dg)\hat{s}+(ah+bg+cf+de)\hat{t} \\ = \left(a+c\hat{r}+b\hat{s}+d\hat{t}\right) \times C_{4.2} \times \left(e+g\hat{r}+f\hat{s}+h\hat{t}\right) \end{pmatrix}$$

Where we have used $\times C_{4.1} \times$ and $\times C_{4.2} \times$ to represent the different multiplication operations within these two algebras – one needs care in doing this.

Multiplication by a scalar:
$$\phi\left(\alpha a + \alpha b\hat{r} + \alpha c\hat{s} + \alpha d\hat{t}\right) \mapsto \left(\alpha a + \alpha d\hat{r} + \alpha c\hat{s} + \alpha b\hat{t}\right)$$

Theorem:

$C_{4.3}$ is isomorphic as an algebra to $C_{4.2}$.

Proof:
$$\phi(C_{4.3} \mapsto C_{4.2}) : \left(a+b\hat{r}+c\hat{s}+d\hat{t} \mapsto a+d\hat{r}+c\hat{s}+b\hat{t}\right)$$

Addition:

$$\phi \begin{pmatrix} \left(a+b\hat{r}+c\hat{s}+d\hat{t}\right)+\left(e+f\hat{r}+g\hat{s}+h\hat{t}\right) \\ =(a+e)+(b+f)\hat{r}+(c+g)\hat{s}+(d+h)\hat{t} \end{pmatrix}$$

$$\mapsto \begin{pmatrix} (a+e)+(d+h)\hat{r}+(c+g)\hat{s}+(b+f)\hat{t} \\ = \left(a+d\hat{r}+c\hat{s}+b\hat{t}\right)+\left(e+h\hat{r}+g\hat{s}+f\hat{t}\right) \end{pmatrix}$$

The Multiplication:
The multiplication operations within these two algebras are different.

$$\phi \begin{pmatrix} \left(a+b\hat{r}+c\hat{s}+d\hat{t}\right) \times C_{4.3} \times \left(e+f\hat{r}+g\hat{s}+h\hat{t}\right) \\ = (ae+bg+cf+dh)+(af+be+ch+dg)\hat{r} \\ +(ag+bh+ce+df)\hat{s}+(ah+bf+cg+de)\hat{t} \end{pmatrix}$$

$$\mapsto \begin{pmatrix} (ae+dg+ch+bf)+(ah+bg+cf+de)\hat{r} \\ +(ag+df+ce+bh)\hat{s}+(af+dh+cg+be)\hat{t} \\ = \left(a+d\hat{r}+c\hat{s}+b\hat{t}\right) \times C_{4.2} \times \left(e+h\hat{r}+g\hat{s}+f\hat{t}\right) \end{pmatrix}$$

Where we have used $\times C_{4.3} \times$ and $\times C_{4.2} \times$ to represent the different multiplication operations within these two algebras – one needs care in doing this.

Multiplication by a scalar:
$$\phi\left(\alpha a + \alpha b\hat{r} + \alpha c\hat{s} + \alpha d\hat{t}\right) \mapsto \left(\alpha a + \alpha d\hat{r} + \alpha c\hat{s} + \alpha b\hat{t}\right)$$

Theorem:

$C_{4.1}$ is isomorphic as an algebra to $C_{4.3}$. *Proof:* Self-evident from above.

Algebraic Matrix Forms by Cayley Tables:

Nomenclature: *Standard Form Cayley Table*
We say that the Cayley table of a group is in Standard Form if it is arranged so the identity appears on the leading diagonal. This is achieved by making the left-most column the ordered inverses of the top-most row.[78]

Consider the Standard Form Cayley Table of the groups $\{C_1, C_2, C_3, C_4, C_2 \times C_2\}$. We list those Standard Form Cayley Tables alongside the Algebraic Matrix Forms $\{C_1, C_2, C_3, C_{4.1}, C_2 \times C_2\}$ within which the space squeezing parameters have been set to unity:

---

[78] The Standard Form of a Cayley table is not unique.

Section 5

$$A \quad : \quad [a] \tag{5.21}$$

$$\begin{matrix} A & B \\ B & A \end{matrix} \quad : \quad \begin{bmatrix} a & b \\ b & a \end{bmatrix}$$

$$\begin{matrix} A & B & C \\ C & A & B \\ B & C & A \end{matrix} \quad : \quad \begin{bmatrix} a & b & c \\ c & a & b \\ b & c & a \end{bmatrix} \tag{5.22}$$

$$\begin{matrix} A & B & C & D \\ D & A & B & C \\ C & D & A & B \\ B & C & D & A \end{matrix} \quad : \quad \begin{bmatrix} a & b & c & d \\ d & a & b & c \\ c & d & a & b \\ b & c & d & a \end{bmatrix} \tag{5.23}$$

$$\begin{matrix} A & B & C & D \\ B & A & D & C \\ C & D & A & B \\ D & C & B & A \end{matrix} \quad : \quad \begin{bmatrix} a & b & c & d \\ b & a & d & c \\ c & d & a & b \\ d & c & b & a \end{bmatrix} \tag{5.24}$$

These algebraic matrix forms can be constructed by simply copying the Standard Form Cayley Tables into a matrix. Of course, one then has to calculate the space squeezing parameters.

In a previous chapter, we discovered how to construct algebraic matrix forms. The first three steps of that construction are:

---
1) Choose the size of the $n \times n$ square matrix.
2) Fill the leading diagonal with a's.
3) Find every possible matrix with $(n-1)$ other independent variables arranged so that each variable occurs once and only once in every column and in every row of the matrix.

Cayley tables in standard form are always square and of a size that is the order of the group. Cayley tables in standard form always have the same element (the identity) on the leading diagonal. Cayley tables in any form always have each element of the group occurring once and only once in each row and in each column of the table.

Complex Numbers The Higher Dimensional Forms – 2nd Edition

Of course, we still have to check for multiplicative commutativity, and the Cartesian form of the algebraic matrix form will require restrictions on the values of the variables.

> **Conjecture:**
> It seems to be the case that abelian groups always produce multiplicatively commutative algebraic matrix forms and that non-abelian groups always produce multiplicatively non-commutative algebraic matrix forms – division algebras. The conjecture is that this is the case for all finite groups.[79]

> 2nd edition note: This conjecture is now known to be wrong. The $C_2 \times C_2 \times ...$ groups produce non-commutative algebras such as the quaternions.

Illegitimate Spaces:

In the first chapter of this section, we chose that all the elements on the leading diagonal must be equal[80] rather than impose restrictions upon the values of the variables. In this chapter, we look at a matrix form that has different variables upon the leading diagonal. We do this because this particular matrix form seems to be the matrix form of the 3-dimensional space we see about us.

Consider:

$$\begin{bmatrix} a & b & 0 \\ -b & a & 0 \\ 0 & 0 & z \end{bmatrix} \qquad (5.25)$$

These matrices are:
- Inclusive of the additive identity
- Inclusive of additive inverses
- Additively closed
- Multiplicatively closed
- Inclusive of multiplicative inverses
- Inclusive of multiplicative identity
- Multiplicatively commutative
- Non-singular if $z \neq 0$ unless $\{a,b\} = 0$

---

[79] 2nd edition note: What a fine example of how wrong we humans can be. The commutative $C_2 \times C_2 \times ...$ groups produce the non-commutative algebras that underpin the whole physics of the universe.

[80] Section 5 - Chapter 1 - Theorem 5

Section 5

With the restrictions, these matrices are an algebraic field.
We split this into two parts:

$$\begin{bmatrix} a & 0 & 0 \\ 0 & a & 0 \\ 0 & 0 & z \end{bmatrix} + \begin{bmatrix} 0 & b & 0 \\ -b & 0 & 0 \\ 0 & 0 & 0 \end{bmatrix} \tag{5.26}$$

We note that these two parts are multiplicatively commutative, and so we can exponentiate this matrix and split it.
We have:

$$\exp\left(\begin{bmatrix} a & 0 & 0 \\ 0 & a & 0 \\ 0 & 0 & z \end{bmatrix} + \begin{bmatrix} 0 & b & 0 \\ -b & 0 & 0 \\ 0 & 0 & 0 \end{bmatrix}\right) = \exp\left(\begin{bmatrix} a & 0 & 0 \\ 0 & a & 0 \\ 0 & 0 & z \end{bmatrix}\right)\exp\left(\begin{bmatrix} 0 & b & 0 \\ -b & 0 & 0 \\ 0 & 0 & 0 \end{bmatrix}\right)$$

$$= \begin{bmatrix} e^a & 0 & 0 \\ 0 & e^a & 0 \\ 0 & 0 & e^z \end{bmatrix} \begin{bmatrix} \cos b & \sin b & 0 \\ -\sin b & \cos b & 0 \\ 0 & 0 & 1 \end{bmatrix}$$

(5.27)

By closure we have:

$$\begin{bmatrix} u & v & 0 \\ -v & u & 0 \\ 0 & 0 & w \end{bmatrix} = \begin{bmatrix} e^a & 0 & 0 \\ 0 & e^a & 0 \\ 0 & 0 & e^z \end{bmatrix} \begin{bmatrix} \cos b & \sin b & 0 \\ -\sin b & \cos b & 0 \\ 0 & 0 & 1 \end{bmatrix} \tag{5.28}$$

The determinants are:

$$\left(u^2 + v^2\right)w = r^2 h \tag{5.29}$$

The right hand matrix is the rotation matrix of the seemingly 3-dimensional space that we see about us. It contains upon its leading diagonal the rotation matrix of 2-dimensional Euclidean space and the rotation matrix of 1-dimensional space. It is thus 2-dimensional Euclidean space together with 1-dimensional space. It is not a bona-fide 3-dimensional space. We note that, when $z = 0$, this is a 2-dimensional space.

The reader might expect that this matrix of everyday 3-dimensional space would give the distance function:

$$d_3^2 = a^2 + b^2 + z^2 \tag{5.30}$$

to which we are accustomed in our everyday lives, but it does not. The understanding that this distance function is the distance function of everyday seemingly 3-dimensional space is in error. The error is the undeclared and incorrect assumption that space is fixed and absolute. We set some axes and calculate a 2-dimensional distance:

$$d_2^2 = a^2 + b^2 \qquad (5.31)$$

We then repeat the procedure to move into 3-dimensions:
$$d_3^2 = d_2^2 + z^2 \qquad (5.32)$$

In both cases, we did a calculation in 2-dimensional space. The illusion that we are calculating a 3-dimensional distance arises from keeping the axes absolutely set. With reference to floors and walls, this works practically, but without the floors and walls to provide an absolute reference frame, we are always calculating in 2-dimensional space.

The product of two such matrices is:

$$\begin{bmatrix} a & b & 0 \\ -b & a & 0 \\ 0 & 0 & z \end{bmatrix} \begin{bmatrix} c & d & 0 \\ -d & c & 0 \\ 0 & 0 & y \end{bmatrix} = \begin{bmatrix} ac-bd & ad+bc & 0 \\ -(ad+bc) & ac-bd & 0 \\ 0 & 0 & yz \end{bmatrix} \qquad (5.33)$$

In non-matrix notation, this algebra is:
$$(a+b\hat{i}+z\hat{j})(c+d\hat{i}+y\hat{j}) = (ac-bd)+(ad+bc)\hat{i}+yz\hat{j} \qquad (5.34)$$

Multiplying the two numbers together distributively and setting $\hat{i}^2 = -1$ gives:
$$(a+b\hat{i}+z\hat{j})(c+d\hat{i}+y\hat{j}) = (ac-bd)+(ad+bc)\hat{i}+(ay+cz)\hat{j}$$
$$+by\hat{i}\hat{j}+dz\hat{j}\hat{i}+yz\hat{j}^2$$
$$(5.35)$$

This is not the algebraic product, and it cannot be made into the algebraic product by setting j in any way. This arises because this algebra is a 2-dimensional algebra and a 1-dimensional algebra fixed together and not a 3-dimensional algebra.[81]

We move into the conventionally accepted form of 4-dimensional space. Consider:

$$\begin{bmatrix} a & b & 0 & 0 \\ -b & a & 0 & 0 \\ 0 & 0 & c & d \\ 0 & 0 & d & c \end{bmatrix} \qquad (5.36)$$

These matrices are:
        Inclusive of the additive identity

---

[81] The unfolded form of this algebra is 5-dimensional – four plus one.

Section 5

Inclusive of additive inverses
Additively closed
Multiplicatively closed
Inclusive of multiplicative inverses
Inclusive of multiplicative identity
Multiplicatively commutative

Non-singular if $c > d$ and $\{a,b\} \neq 0$ unless $\{c,d\} = 0$

With the restrictions, these matrices are an algebraic field.

We split this into two parts:

$$\begin{bmatrix} a & 0 & 0 & 0 \\ 0 & a & 0 & 0 \\ 0 & 0 & c & 0 \\ 0 & 0 & 0 & c \end{bmatrix} + \begin{bmatrix} 0 & b & 0 & 0 \\ -b & 0 & 0 & 0 \\ 0 & 0 & 0 & d \\ 0 & 0 & d & 0 \end{bmatrix} \tag{5.37}$$

We note that these two parts are multiplicatively commutative, and so we can exponentiate this matrix and split it.

We have:

$$\exp\left(\begin{bmatrix} a & 0 & 0 & 0 \\ 0 & a & 0 & 0 \\ 0 & 0 & c & 0 \\ 0 & 0 & 0 & c \end{bmatrix} + \begin{bmatrix} 0 & b & 0 & 0 \\ -b & 0 & 0 & 0 \\ 0 & 0 & 0 & d \\ 0 & 0 & d & 0 \end{bmatrix}\right) \tag{5.38}$$

$$= \begin{bmatrix} e^a & 0 & 0 & 0 \\ 0 & e^a & 0 & 0 \\ 0 & 0 & e^c & 0 \\ 0 & 0 & 0 & e^c \end{bmatrix} \begin{bmatrix} \cos b & \sin b & 0 & 0 \\ -\sin b & \cos b & 0 & 0 \\ 0 & 0 & \cosh d & \sinh d \\ 0 & 0 & \sinh d & \cosh d \end{bmatrix} \tag{5.39}$$

By closure we have:

$$\begin{bmatrix} z & y & 0 & 0 \\ -y & z & 0 & 0 \\ 0 & 0 & x & w \\ 0 & 0 & w & x \end{bmatrix} = \begin{bmatrix} e^a & 0 & 0 & 0 \\ 0 & e^a & 0 & 0 \\ 0 & 0 & e^c & 0 \\ 0 & 0 & 0 & e^c \end{bmatrix} \begin{bmatrix} \cos b & \sin b & 0 & 0 \\ -\sin b & \cos b & 0 & 0 \\ 0 & 0 & \cosh d & \sinh d \\ 0 & 0 & \sinh d & \cosh d \end{bmatrix}$$
(5.40)

The determinants are:

$$(z^2 + y^2)(x^2 - w^2) = r^2 h^2 \tag{5.41}$$

The right hand matrix is the rotation matrix of the 4-dimensional space that is conventionally called space-time. It contains upon its leading diagonal the rotation matrix of 2-dimensional Euclidean space and the rotation matrix of 2-dimensional space-time. It is thus two 2-dimensional spaces stuck together. It is not a bona-fide 4-dimensional space.

Again, the distance function is not what the reader might have expected. Again, the expected distance function is founded in the undeclared assumption that space is absolute (axes are fixed is the same thing).

The special theory of relativity, however it is derived, is a theory set in a 2-dimensional space-time. The addition of the two extra dimensions is a 'fudge'.

2nd edition note: We now know that our space-time with the correct distance function emerges as an emergent expectation space from the superimposition of the six non-commutative A_3 algebras that are within the $C_2 \times C_2$ group. See: Dennis Morris Upon General Relativity.

Section 6

2nd edition note: Now-a-days, we do not worry about formal definitions of trigonometric functions as we did in the 1st edition of this book. We have become used to the higher dimensional trigonometric functions to the point where we simply observe them to exist. This chapter reflects your author's initial reluctance to accept such a change in the views with which he was inculcated at school.

Introduction to Section 6:

Each of the 1-dimensional and 2-dimensional natural algebras has within it a space. The real numbers hold within them the 1-dimensional space; the Euclidean complex numbers contain the 2-dimensional Euclidean space, and the hyperbolic complex numbers contain the 2-dimensional hyperbolic space. We have found other natural algebras, but do they contain a space and, if so, how will we recognize it as a space? What mathematical 'objects' need be on our page before we can proclaim that what we have is a geometric space?

In this section, we seek to define what we mean by geometric space. We do this by listing the mathematical 'things', and the properties of those 'things' that we feel are sufficient and necessary to make geometric space.

When we have the mathematical objects that constitute a geometric space, it does not mean the geometric space will manifest itself to our senses as geometric space. How it appears to us is a matter of psychology not mathematics.

Part of this section (addendum 1 to chapter 3) is a lengthy formal definition of what we mean by a trigonometric function. There is much in that definition that we do not meet until later in this work. The definition is included in this section because this is the proper place for it, but the reader is advised to skip the reading of it in the first reading of this work.

The Definition of the Trigonometric Functions:

In 2-dimensions, there are two types of flat space associated with the natural algebras. There is the 2-dimensional Euclidean space associated with the Euclidean complex numbers, and there is the 2-dimensional hyperbolic space associated with the hyperbolic complex numbers. Associated with each of these spaces is a set of trigonometric functions. These are the Euclidean trigonometric functions, $\{\cos, \sin\}$, together with the various other 'mixtures' of these two functions, and the hyperbolic trigonometric functions, $\{\cosh, \sinh\}$, together with the various other 'mixtures' of these two functions. What is it about these functions that justifies the use of the adjective 'trigonometric'? What properties is a function required to have, or what need it be, before it can be called a trigonometric function?

The 2-dimensional Euclidean trigonometric functions, $\{\cos, \sin\}$, were first defined in 6th century India. They were defined to be the projections from a point on the unit circle on to each of the axes of the Euclidean plane. This has been the standard definition used by geometers since then. In the case of 2-dimensional Euclidean geometry, the cosine is the projection from the unit circle on to the horizontal axis and the sine is the projection from the unit circle on to the vertical axis. In the case of 2-dimensional hyperbolic geometry, the *cosh* is the projection from the hyperbolic unit circle on to the horizontal axis and the *sinh* is the projection from the hyperbolic unit circle on to the vertical axis. Of course, if we draw the hyperbolic unit circle of 2-dimensional hyperbolic space on 2-dimensional Euclidean paper, it appears as a hyperbola.

This definition does not fit 3-dimensional space automatically. We can easily extend the unit circle to be a unit sphere (and analogously so for higher dimensional spaces), but a projection parallel to an axis from a point on the 3-dimensional unit sphere might fall on to a plane formed by the other two axes and not on to a particular axis. Such a projection is a point. If we want to project from the 3-dimensional unit sphere in a direction parallel to an axis on to an axis, we need to first project down to a point on a plane and then project down out of that plane to a point on the axis. Putting these two projections together forms a function of a function, and this function of a function will project down from a point on the unit sphere to a point on an axis. In general, the projection is not parallel to a particular axis. In a 4-dimensional space, we would require to do three axis parallel projections, which is a function of a function of a function, to arrive at an axis.

Nomenclature:
Simple Trigonometric Function
A projection function that is a single projection from the unit sphere is called a simple trigonometric function.

Compound Trigonometric Function

Section 6

> A projection function that is a projection of a projection of a ... from the unit sphere on to an axis of the space is called a compound trigonometric function.

We proceed with a prospective geometric definition of a trigonometric function.

> Prospective Geometric Definition:
> *Trigonometric Function*
> A trigonometric function is a projection function from the *n*-dimensional unit sphere on to an axis – a compound trigonometric function.

Clearly, every 3-dimensional space has three axes and hence will have associated with it a set of three compound trigonometric functions that are projections on to axes. Clearly, every 3-dimensional space also has three planes and hence will have associated with it another set of three simple trigonometric functions that are projections on to planes.

The above definition catches the essential geometric properties of the trigonometric functions $\{\cos, \sin, \cosh, \sinh\}$. However, these functions have other aspects to them. They appear in the rotation matrices of the space with which they are associated. Indeed, the elements of the rotation matrices are all trigonometric functions. We seek to capture this property of the trigonometric functions with a prospective rotation definition.

> Prospective Rotation Definition:
> *Trigonometric Function*
> A trigonometric function is a function that appears in a rotation matrix.

Of course, if the space does not have a rotation matrix, it cannot have any trigonometric functions. This does not concern us for what kind of a space is it in which one cannot rotate?[82]

In fact, these definitions are not different. If we rotate the position matrix $\begin{bmatrix} 1 & 0 \\ 0 & 1 \end{bmatrix}$, the new position matrix will have, as elements, projections on to the axes from the unit sphere (circle). If the rotation matrix did not contain such projection functions, it could not be a rotation matrix.

The trigonometric functions have another property that we seek to capture in our definition. In 2-dimensions, the rate of change of the projection on to one

---

[82] The 1-dimensional one, perhaps.

axis with respect to the change in angle from that axis is the projection on to the other axis. In the case of the 2-dimensional hyperbolic functions, we have:

$$\left\{\frac{d(\cosh x)}{dx} = \sinh x, \frac{d(\sinh x)}{dx} = \cosh x\right\} \quad (6.1)$$

with similar, slightly more complicated, relations for the 2-dimensional Euclidean functions. In 2-dimensions, the trigonometric functions have series expansions that are splittings of the exponential series (in the Euclidean case, with some minus signs chucked in). In the case of the 2-dimensional hyperbolic functions, we have:

$$\cosh x = \frac{x^0}{0!} + \frac{x^2}{2!} + \frac{x^4}{4!} + \ldots$$
$$\sinh x = \frac{x^1}{1!} + \frac{x^3}{3!} + \frac{x^5}{5!} + \ldots \quad (6.2)$$

The essence is that a set of trigonometric functions has a differentiation cycle. Thus we have a prospective calculus definition of a trigonometric function.

Prospective Calculus Definition:
Trigonometric Function
A set of trigonometric functions is a set of functions that have a differentiation cycle.

Finally, we decide to cheat. We define a trigonometric function to be that which is an element of the matrices with determinant unity that are produced when a natural algebra is exponentiated. It turns out that this definition satisfies all the previous prospective definitions.

Cheating Definition:
Trigonometric Function
A trigonometric function is an element of the matrices with determinant unity that are produced when a natural algebra is exponentiated.

However, hindsight allows us to give a precise analytical form of this definition. This precise definition is given as an addendum to this chapter, and this precise definition is the definition that we take to be the proper definition of a trigonometric function. This precise definition satisfies all the prospective definitions given above. The functions it defines are all projection functions that appear in rotation matrices, that have differentiation cycles, and that drop out of the natural algebras.

Section 6

| Precise Definition: | *Trigonometric Function* | See Addendum. |

Addendum 1

Definition of Simple Trigonometric Functions:[83]

We first deal with the functions that are elements of the rotation matrices that occur in the simple polar form of algebraic matrix forms. These are the simple trigonometric functions.

The basic trigonometric functions of 2-dimensional space have series expansions that are 2-way splittings of the exponential series. These splittings of the exponential series have hypergeometric representation.

$$\cos x = hypergeom\left([\ ],\left[\frac{1}{2}\right],-\left(\frac{x}{2}\right)^2\right) = \frac{x^0}{0!} - \frac{x^2}{2!} + \frac{x^4}{4!} - \frac{x^6}{6!} + \ldots$$

$$\sin x = x\left(hypergeom\left([\ ],\left[\frac{3}{2}\right],-\left(\frac{x}{2}\right)^2\right)\right) = \frac{x^1}{1!} - \frac{x^3}{3!} + \frac{x^5}{5!} - \frac{x^7}{7!} + \ldots$$

$$\cosh x = hypergeom\left([\ ],\left[\frac{1}{2}\right],\left(\frac{x}{2}\right)^2\right) = \frac{x^0}{0!} + \frac{x^2}{2!} + \frac{x^4}{4!} + \frac{x^6}{6!} + \ldots$$

$$\sinh x = x\left(hypergeom\left([\ ],\left[\frac{3}{2}\right],\left(\frac{x}{2}\right)^2\right)\right) = \frac{x^1}{1!} + \frac{x^3}{3!} + \frac{x^5}{5!} + \frac{x^7}{7!} + \ldots$$

(6.3)

The elements in the rotation matrices of the simple polar form of the 3-dimensional algebraic matrix forms are 3-way splittings of the exponential series with hypergeometric representation:

$$AH_3(x) = hypergeom\left([\ ],\left[\frac{1}{3},\frac{2}{3}\right],\left(\frac{x}{3}\right)^3\right) = \frac{x^0}{0!} + \frac{x^3}{3!} + \frac{x^6}{6!} + \frac{x^9}{9!} + \ldots$$

$$BH_3(x) = \frac{x}{1!}\left(hypergeom\left([\ ],\left[\frac{2}{3},\frac{4}{3}\right],\left(\frac{x}{3}\right)^3\right)\right) = \frac{x^1}{1!} + \frac{x^4}{4!} + \frac{x^7}{7!} + \frac{x^{10}}{10!} + \ldots$$

$$CH_3(x) = \frac{x^2}{2!}\left(hypergeom\left([\ ],\left[\frac{4}{3},\frac{5}{3}\right],\left(\frac{x}{3}\right)^3\right)\right) = \frac{x^2}{2!} + \frac{x^5}{5!} + \frac{x^8}{8!} + \frac{x^{11}}{11!} + \ldots$$

(6.4)

---

[83] At a first reading, much of this material will be meaningless to the reader. We are dealing with things that are not revealed until later in this work. The reader is urged to skip through these definitions and proceed to the next chapter.

$$AE_3(x) = \text{hypergeom}\left([\ ], \left[\frac{1}{3}, \frac{2}{3}\right], -\left(\frac{x}{3}\right)^3\right) = \frac{x^0}{0!} - \frac{x^3}{3!} + \frac{x^6}{6!} - \frac{x^9}{9!} + \ldots$$

$$BE_3(x) = \frac{x}{1!}\left(\text{hypergeom}\left([\ ], \left[\frac{2}{3}, \frac{4}{3}\right], -\left(\frac{x}{3}\right)^3\right)\right) = \frac{x^1}{1!} - \frac{x^4}{4!} + \frac{x^7}{7!} - \frac{x^{10}}{10!} + \ldots$$

$$CE_3(x) = \frac{x^2}{2!}\left(\text{hypergeom}\left([\ ], \left[\frac{4}{3}, \frac{5}{3}\right], -\left(\frac{x}{3}\right)^3\right)\right) = \frac{x^2}{2!} - \frac{x^5}{5!} + \frac{x^8}{8!} - \frac{x^{11}}{11!} + \ldots$$

(6.5)

The elements in the rotation matrices of the simple polar form of the 4-dimensional algebraic matrix forms are 4-way splittings of the exponential series with hypergeometric representation:

$$AH_4(x) = \text{hypergeom}\left([\ ], \left[\frac{1}{4}, \frac{1}{2}, \frac{3}{4}\right], \left(\frac{x}{4}\right)^4\right) = 1 + \frac{x^4}{4!} + \frac{x^8}{8!} + \frac{x^{12}}{12!} + \ldots$$

$$BH_4(x) = x\left(\text{hypergeom}\left([\ ], \left[\frac{1}{2}, \frac{3}{4}, \frac{5}{4}\right], \left(\frac{x}{4}\right)^4\right)\right) = x + \frac{x^5}{5!} + \frac{x^9}{9!} + \frac{x^{13}}{13!} + \ldots$$

$$CH_4(x) = \frac{x^2}{2}\left(\text{hypergeom}\left([\ ], \left[\frac{3}{4}, \frac{5}{4}, \frac{6}{4}\right], \left(\frac{x}{4}\right)^4\right)\right) = \frac{x^2}{2!} + \frac{x^6}{6!} + \frac{x^{10}}{10!} + \ldots$$

$$DH_4(x) = \frac{x^3}{6}\left(\text{hypergeom}\left([\ ], \left[\frac{5}{4}, \frac{6}{4}, \frac{7}{4}\right], \left(\frac{x}{4}\right)^4\right)\right) = \frac{x^3}{3!} + \frac{x^7}{7!} + \frac{x^{11}}{11!} + \ldots$$

(6.6)

$$AE_4(x) = \text{hypergeom}\left([\ ], \left[\frac{1}{4}, \frac{1}{2}, \frac{3}{4}\right], -\left(\frac{x}{4}\right)^4\right) = 1 - \frac{x^4}{4!} + \frac{x^8}{8!} - \frac{x^{12}}{12!} + \ldots$$

$$BE_4(x) = x\left(\text{hypergeom}\left([\ ], \left[\frac{1}{2}, \frac{3}{4}, \frac{5}{4}\right], -\left(\frac{x}{4}\right)^4\right)\right) = x - \frac{x^5}{5!} + \frac{x^9}{9!} - \frac{x^{13}}{13!} + \ldots$$

$$CE_4(x) = \frac{x^2}{2}\left(\text{hypergeom}\left([\ ], \left[\frac{3}{4}, \frac{5}{4}, \frac{6}{4}\right], -\left(\frac{x}{4}\right)^4\right)\right) = \frac{x^2}{2!} - \frac{x^6}{6!} + \frac{x^{10}}{10!} - \ldots$$

$$DE_4(x) = \frac{x^3}{6}\left(\text{hypergeom}\left([\ ], \left[\frac{5}{4}, \frac{6}{4}, \frac{7}{4}\right], -\left(\frac{x}{4}\right)^4\right)\right) = \frac{x^3}{3!} - \frac{x^7}{7!} + \frac{x^{11}}{11!} - \ldots$$

(6.7)

The pattern within the hypergeometric format is clear. By similar means, we can define the simple trigonometric function for *n*-dimensions as *n*-ways splittings of the exponential.

Section 6

Definition:
Simple Trigonometric Function
The hypergeometric representation of the H-type functions is always of the form:

$$AH_n = \frac{x^0}{0!}\left(\text{hypergeom}\left([\ ],\left[\frac{1}{n},\frac{2}{n},...,\frac{n-1}{n}\right],\left(\frac{x}{n}\right)^n\right)\right)$$

$$= \frac{x^0}{0!} + \frac{x^n}{n!} + ...$$

$$BH_n = \frac{x^1}{1!}\left(\text{hypergeom}\left([\ ],\left[\frac{2}{n},\frac{3}{n},...,\frac{n-1}{n},\frac{n+1}{n}\right],\left(\frac{x}{n}\right)^n\right)\right)$$

$$= \frac{x^1}{1!} + \frac{x^{n+1}}{(n+1)!} + ... \qquad (6.8)$$

...

$$\Theta H_n = \frac{x^{n-1}}{(n-1)!}\left(\text{hypergeom}\left([\ ],\left[\frac{n+1}{n},\frac{n+2}{n},...,\frac{2n-1}{n}\right],\left(\frac{x}{n}\right)^n\right)\right)$$

$$= \frac{x^{n-1}}{(n-1)!} + \frac{x^{2n-1}}{(2n-1)!} + ...$$

where ΘH_n is the *n*-ways splitting of the exponential series whose first term is x^{n-1}. Note that the term $\frac{n}{n}$ is always absent from the middle bracket.

The hypergeometric representation of the E-type functions is always of the form:

$$AE_n = \frac{x^0}{0!}\left(\text{hypergeom}\left([\ ],\left[\frac{1}{n},\frac{2}{n},...,\frac{n-1}{n}\right],-\left(\frac{x}{n}\right)^n\right)\right)$$

$$(6.9)$$

$$= \frac{x^0}{0!} - \frac{x^n}{n!} + ...$$

$$BE_n = \frac{x^1}{1!}\left(\text{hypergeom}\left([\ ],\left[\frac{2}{n},\frac{3}{n},...,\frac{n-1}{n},\frac{n+1}{n}\right],-\left(\frac{x}{n}\right)^n\right)\right)$$

$$(6.10)$$

$$= \frac{x^1}{1!} - \frac{x^{n+1}}{(n+1)!} + ...$$

$$\Theta E_n = \frac{x^{n-1}}{(n-1)!}\left(hypergeom\left([\ ],\left[\frac{n+1}{n},\frac{n+2}{n},....,\frac{2n-1}{n}\right],-\left(\frac{x}{n}\right)^n\right)\right) \quad (6.11)$$

$$= \frac{x^{n-1}}{(n-1)!} - \frac{x^{2n-1}}{(2n-1)!} + ...$$

Of course:

$$hypergeom\left([\ ],[\ ],\left(\frac{x}{1}\right)^1\right) = e^x \quad \& \quad hypergeom\left([\ ],[\ ],-\left(\frac{x}{1}\right)^1\right) = e^{-x}$$
(6.12)

We do not need the hypergeometric form of these functions; it is included to give precision to our definition and to emphasize the similar nature of these functions to the 2-dimensional trigonometric functions.

Notes:
We always have:

$$AE_n = AH_{2n} - \Theta H_{2n} \quad (6.13)$$

where ΘH_{2n} is the $2n$-ways splitting of the exponential series whose first term is x^{2n-1}, with similar relations for $BE_n, CE_n,...$ We always have:

$$AH_n = AH_{2n} + \Theta H_{2n} \quad (6.14)$$

where ΘH_{2n} is the $2n$-ways splitting of the exponential series whose first term is x^{2n-1}, with similar relations for $BH_n, CH_n,...$ Similar relations can be derived (think of the series expansions) from the addition of more than two simple-trig functions, for example:

$$AH_3 = AH_9 + DH_9 + GH_9 \quad (6.15)$$

So it is that the higher dimensional simple-trig functions are 'folded' inside the lower dimensional simple-trig functions. We always have:

$$XH_{2n} = \frac{1}{2}(XH_n + XE_n) \quad (6.16)$$

where X represents any of the simple-trig functions. So it is that once we have established both the H-type and the E-type simple-trig functions for a particular n, the higher dimensional H-type simple-trig functions of dimension $2n$, $4n$, $8n$ ... are determined. Of course, the H-type simple-trig functions of dimension n where n is odd are not so determined. Re-arranging this equation gives:

Section 6

$$XE_n = 2XH_{2n} - XH_n \qquad (6.17)$$

So it is that once we have established the H-type simple-trig functions for a particular n and $2n$, the n-dimensional E-type functions are determined.

Definition of Compound Trigonometric Functions:[84]

(We will also use the term v-functions to mean compound trigonometric functions.)

We now deal with the functions that are elements of the single compound rotation matrix that occurs in the compound polar form of algebraic matrix forms. These are the compound trigonometric functions.

The compound trigonometric functions are projections of projections of .... They are simple trigonometric functions of simple trigonometric functions of ... They occur in the rotation matrix that is the product of the simple rotation matrices in the simple polar form of an algebraic matrix form. An example: The simple polar form of the 3-dimensional algebraic matrix form, $C_3 I_3^1 H^2_{(j=1,k=1)}$, is:

$$\exp\left(\begin{bmatrix} a & b & c \\ c & a & b \\ b & c & a \end{bmatrix}\right) = PROD \left\{ \begin{bmatrix} r & 0 & 0 \\ 0 & r & 0 \\ 0 & 0 & r \end{bmatrix} \begin{bmatrix} AH_3(b) & BH_3(b) & CH_3(b) \\ CH_3(b) & AH_3(b) & BH_3(b) \\ BH_3(b) & CH_3(b) & AH_3(b) \end{bmatrix} \begin{bmatrix} AH_3(c) & CH_3(c) & BH_3(c) \\ BH_3(c) & AH_3(c) & CH_3(c) \\ CH_3(c) & BH_3(c) & AH_3(c) \end{bmatrix} \right\}$$

(6.18)

We set:

$$v_{\left[C_3 I_3 H^2_{(j=1,k=1)}\right]} A(b,c) = AH_3(b) AH_3(c) + BH_3(b) BH_3(c) + CH_3(b) CH_3(c)$$

$$v_{\left[C_3 I_3 H^2_{(j=1,k=1)}\right]} B(b,c) = AH_3(c) BH_3(b) + AH_3(b) CH_3(c) + BH_3(c) CH_3(b)$$

$$v_{\left[C_3 I_3 H^2_{(j=1,k=1)}\right]} C(b,c) = AH_3(c) CH_3(b) + AH_3(b) BH_3(c) + BH_3(b) CH_3(c)$$

(6.19)

---

[84] At a first reading, much of this material will be meaningless to the reader. We are dealing with things that are not revealed until later in this work. The reader is urged to skip through these definitions and proceed to the next chapter.

Leading to the compound polar form:

$$\exp\left(\begin{bmatrix} a & b & c \\ c & a & b \\ b & c & a \end{bmatrix}\right) =$$

$$\begin{bmatrix} r & 0 & 0 \\ 0 & r & 0 \\ 0 & 0 & r \end{bmatrix} \begin{bmatrix} v_{[C,L^1H^1_{(\ldots)}]}A(b,c) & v_{[C,L^1H^1_{(\ldots)}]}B(b,c) & v_{[C,L^1H^1_{(\ldots)}]}C(b,c) \\ v_{[C,L^1H^1_{(\ldots)}]}C(b,c) & v_{[C,L^1H^1_{(\ldots)}]}A(b,c) & v_{[C,L^1H^1_{(\ldots)}]}B(b,c) \\ v_{[C,L^1H^1_{(\ldots)}]}B(b,c) & v_{[C,L^1H^1_{(\ldots)}]}C(b,c) & v_{[C,L^1H^1_{(\ldots)}]}A(b,c) \end{bmatrix}$$

(6.20)

The functions in the rotation matrix are the compound trigonometric functions of this particular algebra. We call them the v-functions.[85]

The Definition of Rotation:

In 2-dimensions, we have two types of rotation, Euclidean rotation and hyperbolic rotation. Euclidean rotation is a multi-valued thing in that $\{0, 2\pi, 4\pi, \ldots\}$ are all the same angle. Nothing like this occurs in hyperbolic rotation. We seek to define 'rotation' and 'rotation matrix' in such a way that includes everything that these two rotations have in common but excludes everything that they do not have in common.

A rotation is a movement in space. As such it is a linear transformation. As such it can be represented as a matrix. The properties of a rotation are:
i) The distance of a position (as calculated using the appropriate distance function) from the origin is the same after the rotation as it was before the rotation. Technically speaking, we say that the rotational linear transformation leaves the distance function invariant.
ii) Two successive rotations through angles α and β are the same as a single rotation through the angle $(\alpha + \beta)$.
iii) The elements within a rotation matrix are trigonometric functions.

---

[85] They are a recent discovery.

Section 6

In general, the distance functions of the *n*-dimensional natural algebraic spaces are the nth roots[86] of the determinants of the matrix representations of the associated *n*-dimensional natural algebras. The requirement that a rotation matrix leaves the metric invariant is thus the requirement that the rotation matrix has determinant equal to unity.

$Rot(\alpha + \beta) = Rot(\alpha) Rot \beta$ means that rotation is of an exponential nature - $p^{a+b} = p^a p^b$, and we will have to use exponentiation to construct the rotation matrices.

With this understanding, we write our definition of rotation.

Definition:

Rotation

A rotation matrix, *A*, is such that:

i) Multiplication by it leaves the metric invariant. That is: $\det(A) = 1$.

ii) $A(\alpha) \bullet A(\beta) = A(\alpha + \beta)$

iii) The elements of *A* are all trigonometric functions.

The requirement that the elements within a rotation matrix are trigonometric functions and the requirement that the determinant of a rotation matrix be unity will automatically lead to trigonometric identities similar to:

$$\det\left(\begin{bmatrix} \cos\alpha & \sin\alpha \\ -\sin\alpha & \cos\alpha \end{bmatrix}\right) = \cos^2\alpha + \sin^2\alpha = 1 \qquad (6.21)$$

Constructing the rotation matrices by exponentiation will automatically lead to multiple angle identities similar to:

$$\exp\left(\begin{bmatrix} 0 & b+c \\ -(b+c) & 0 \end{bmatrix}\right) = \begin{bmatrix} \cos(b+c) & \sin(b+c) \\ -\sin(b+c) & \cos(b+c) \end{bmatrix} \qquad (6.22)$$

$$\exp\left(\begin{bmatrix} 0 & b+c \\ -(b+c) & 0 \end{bmatrix}\right) = \exp\left(\begin{bmatrix} 0 & b \\ -b & 0 \end{bmatrix}\right) \exp\left(\begin{bmatrix} 0 & c \\ -c & 0 \end{bmatrix}\right) \qquad (6.23)$$

---

[86] We have not yet approached this aspect of the algebras in more than two dimensions.

$$= \begin{bmatrix} \cos b & \sin b \\ -\sin b & \cos b \end{bmatrix} \begin{bmatrix} \cos c & \sin c \\ -\sin c & \cos c \end{bmatrix}$$

$$= \begin{bmatrix} \cos b \cos c - \sin b \sin c & \cos b \sin c + \sin b \cos c \\ \sim & \sim \end{bmatrix} \quad (6.24)$$

$$\cos(b+c) = \cos b \cos c - \sin b \sin c$$
$$\sin(b+c) = \cos b \sin c + \sin b \cos c$$

We conclude this chapter with a look at some rotation matrices. The 2-dimensional Euclidean rotation matrix is:

$$\begin{bmatrix} \cos \alpha & \sin \alpha \\ -\sin \alpha & \cos \alpha \end{bmatrix} \quad (6.25)$$

The 2-dimensional hyperbolic rotation matrix is:

$$\begin{bmatrix} \cosh \alpha & \sinh \alpha \\ \sinh \alpha & \cosh \alpha \end{bmatrix} \quad (6.26)$$

The rotation matrix of 1-dimensional space is:

$$[1] \quad (6.27)$$

The rotation matrix of three 1-dimensional spaces fixed together[87] – that is a conventional 3-dimensional space – is:

$$\begin{bmatrix} 1 & 0 & 0 \\ 0 & 1 & 0 \\ 0 & 0 & 1 \end{bmatrix} \quad (6.28)$$

Of course, there is no concept of angle in 1-dimensional space, and hence the functions within the rotation matrix take no arguments. There is no rotation in 1-dimensional space.

The rotation matrix of 2-dimensional Euclidean space fixed to the 1-dimensional space is:

$$\begin{bmatrix} \cos \alpha & \sin \beta & 0 \\ \sin \beta & \cos \alpha & 0 \\ 0 & 0 & 1 \end{bmatrix} \quad (6.29)$$

This is the rotation matrix of perceived spatial reality. Notice how the 1 does not take α as an argument. This is rotation in a 2-dimensional plane.

---

[87] The real number 1 is the 1-dimensional trigonometric function.

Section 6

The rotation matrix of perceived 4-dimensional reality is:

$$\begin{bmatrix} \cos a & \sin a & 0 & 0 \\ -\sin a & \cos a & 0 & 0 \\ 0 & 0 & \cosh b & \sinh b \\ 0 & 0 & \sinh b & \cosh b \end{bmatrix} \quad (6.30)$$

Of course, hyperbolic angles are different things from Euclidean angles and so this matrix takes two different types of argument. This is not a 4-dimensional space; it is two 2-dimensional spaces (with two different types of angle) fixed together.

We jump the gun to show the reader a simple rotation matrix of a natural 3-dimensional space. A simple rotation matrix of the $C_3 L^1 H^2_{(j=1,k=1)}$ space is:

$$\begin{bmatrix} AH_3(b) & BH_3(b) & CH_3(b) \\ CH_3(b) & AH_3(b) & BH_3(b) \\ BH_3(b) & CH_3(b) & AH_3(b) \end{bmatrix} \quad (6.31)$$

where the functions within the matrix are 3-dimensional simple trigonometric functions.

The compound rotation matrix of this space is:

$$\begin{bmatrix} V_{[C_3 L^1 H^2_{(j=1,k=1)}]} A(b,c) & V_{[C_3 L^1 H^2_{(j=1,k=1)}]} B(b,c) & V_{[C_3 L^1 H^2_{(j=1,k=1)}]} C(b,c) \\ V_{[C_3 L^1 H^2_{(j=1,k=1)}]} C(b,c) & V_{[C_3 L^1 H^2_{(j=1,k=1)}]} A(b,c) & V_{[C_3 L^1 H^2_{(j=1,k=1)}]} B(b,c) \\ V_{[C_3 L^1 H^2_{(j=1,k=1)}]} B(b,c) & V_{[C_3 L^1 H^2_{(j=1,k=1)}]} C(b,c) & V_{[C_3 L^1 H^2_{(j=1,k=1)}]} A(b,c) \end{bmatrix}$$
(6.32)

The compound trigonometric functions take two arguments.

The Definition of the Distance Function:

We would like to call our distance functions metrics, but that word is already in use with a precise meaning that is different from what we mean by distance function. Although some of the distance functions we will meet satisfy the axioms of a metric, not all of them will. Since the 2-dimensional hyperbolic space-time of special relativity, and hence of reality, does not satisfy all the metric axioms, this is not necessarily a great loss.

> Definition:
> ### Distance Function
> The distance function of a natural space is the nth root of the determinant of the associated algebraic matrix form.

This definition is forced upon us by the polar forms of the algebras. It is quite different from the spaces to which we are accustomed.

The association between the determinant of a matrix and space is not new to mathematics. Conventionally, the determinant is seen as the n-dimensional volume of the parallelepiped formed by the vectors within the matrix. If the vectors are orthogonal, we have a n-dimensional square parallelepiped and the length of a side is the nth root of the determinant.

The Definition of Natural Space:

There's that propaganda word 'natural' again. Remember, the algebras we are dealing with are only natural in that they are matrices with matrix multiplication rather than some other form of multiplication.

We define natural space to be a particular bundle of mathematical objects. This bundle includes a distance function, a set of rotation matrices, and a set of trigonometric functions. Nothing less than this will suffice, and we contend that nothing more is required to make a natural geometric space.

We require the trigonometric functions to be the ones that appear in the rotation matrices. We require the distance function to have as many variables as there are different trigonometric functions within the set of compound (or simple) trigonometric functions; this means as many axes as there are projections on to those axes. We require that the distance function declare itself to be a distance function rather than us have to impose this nature upon it.

> Definition:
> ### n-dimensional Natural Space
> i) A Distance Function
>
> ii) A set of $(n-1)$ Simple Rotation Matrices
>
> iii) A set of n Simple Trigonometric Functions
>
> Note: The set of simple rotation matrices will generate a compound rotation matrix.

> Note: The set of simple trigonometric functions will generate a set of compound trigonometric functions.

By an amazing co-incidence, the above mathematical objects are exactly what fall out of the natural algebras. People who do not know me will suspect that I have rigged it. The reader should now consider whether or not the above definition is a 'fair' definition of a geometric space.

These spaces might not manifest themselves to us as space. We require a mind to lay an imaginary blanket over the top of these mathematical objects.

> 2nd edition note: We now have a much clearer understanding of how the spinor algebras manifest themselves, through superimposition of isomorphic algebras, as observed space-time. We understand why our 4-dimensional space holds two types of 2-dimensional rotations rather than one 4-dimensional rotation, why it can be curved and why that curvature is of a 2-dimensional nature, why we have 2-dimensional local flatness and much more. See: Dennis Morris Upon General Relativity & Dennis Morris The Physics of Empty Space.

Section 7

Introduction to Section 7:

In this section, we meet the 3-dimensional natural algebras. With them we meet the 3-dimensional natural spaces together with the 3-dimensional trigonometric functions. These 3-dimensional spaces are nothing like the 3-dimensional space to which we are accustomed. The 3-dimensional trigonometric functions are likely to be new to the first time reader who might have thought that there were only four basic trigonometric functions $\{\sin(\ ), \cos(\ ), \sinh(\ ), \cosh(\ )\}$. We also meet the concepts of shadow algebras and shadow spaces. Much of what is in this section is nothing more than the generalisation of the 2-dimensional natural algebras with which we are reasonably familiar.

The Polar Form of the $C_3 L^1 H^2{}_{(j=1,k=1)}$ Algebra:

We have previously established that the only form of 3×3 matrices with three independent variables that is an algebraic field is (with restrictions):

$$C_3 = \begin{bmatrix} a & b & c \\ jc & a & kb \\ jb & \left(\dfrac{j}{k}\right)c & a \end{bmatrix} : j \in \mathbb{R}, k \in \mathbb{R} \neq 0 \quad (7.1)$$

With $\{j,k\} = \pm 1$, the only flat forms of 3×3 matrices (with three independent variables) that are an algebraic field are (with restrictions):

$$\left\{ C_3 L^1 H^2{}_{(j=1,k=1)} = \begin{bmatrix} a & b & c \\ c & a & b \\ b & c & a \end{bmatrix}, C_3 L^1 E^1 H^1{}_{(j=-1,k=1)} = \begin{bmatrix} a & b & c \\ -c & a & b \\ -b & -c & a \end{bmatrix}, \right.$$

$$\left. C_3 L^1 E^2{}_{(j=1,k=-1)} = \begin{bmatrix} a & b & c \\ c & a & -b \\ b & -c & a \end{bmatrix}, C_3 L^1 E^1 H^1{}_{(j=-1,k=-1)} = \begin{bmatrix} a & b & c \\ -c & a & -b \\ -b & c & a \end{bmatrix} \right\}$$

(7.2)

We have:

Section 7

$$\exp\left(\begin{bmatrix} a & b & c \\ c & a & b \\ b & c & a \end{bmatrix}\right)$$

$$= \exp\left(\begin{bmatrix} a & 0 & 0 \\ 0 & a & 0 \\ 0 & 0 & a \end{bmatrix} + \begin{bmatrix} 0 & b & 0 \\ 0 & 0 & b \\ b & 0 & 0 \end{bmatrix} + \begin{bmatrix} 0 & 0 & c \\ c & 0 & 0 \\ 0 & c & 0 \end{bmatrix}\right)$$

$$= \exp\left(\begin{bmatrix} a & 0 & 0 \\ 0 & a & 0 \\ 0 & 0 & a \end{bmatrix}\right) \exp\left(\begin{bmatrix} 0 & b & 0 \\ 0 & 0 & b \\ b & 0 & 0 \end{bmatrix}\right) \exp\left(\begin{bmatrix} 0 & 0 & c \\ c & 0 & 0 \\ 0 & c & 0 \end{bmatrix}\right)$$

$$= \text{PROD} \begin{pmatrix} \begin{bmatrix} e^a & 0 & 0 \\ 0 & e^a & 0 \\ 0 & 0 & e^a \end{bmatrix} \\ \begin{bmatrix} 1 + \frac{b^3}{3!} + \frac{b^6}{6!} + \ldots & b + \frac{b^4}{4!} + \frac{b^7}{7!} + \ldots & \frac{b^2}{2!} + \frac{b^5}{5!} + \ldots \\ \frac{b^2}{2!} + \frac{b^5}{5!} + \ldots & 1 + \frac{b^3}{3!} + \frac{b^6}{6!} + \ldots & b + \frac{b^4}{4!} + \frac{b^7}{7!} + \ldots \\ b + \frac{b^4}{4!} + \frac{b^7}{7!} + \ldots & \frac{b^2}{2!} + \frac{b^5}{5!} + \ldots & 1 + \frac{b^3}{3!} + \frac{b^6}{6!} + \ldots \end{bmatrix} \\ \begin{bmatrix} 1 + \frac{c^3}{3!} + \frac{c^6}{6!} + \ldots & \frac{c^2}{2!} + \frac{c^5}{5!} + \ldots & c + \frac{c^4}{4!} + \frac{c^7}{7!} + \ldots \\ c + \frac{c^4}{4!} + \frac{c^7}{7!} + \ldots & 1 + \frac{c^3}{3!} + \frac{c^6}{6!} + \ldots & \frac{c^2}{2!} + \frac{c^5}{5!} + \ldots \\ \frac{c^2}{2!} + \frac{c^5}{5!} + \ldots & c + \frac{c^4}{4!} + \frac{c^7}{7!} + \ldots & 1 + \frac{c^3}{3!} + \frac{c^6}{6!} + \ldots \end{bmatrix} \end{pmatrix} \quad (7.3)$$

The infinite series within the second two matrices are 3-way splittings of the exponential series. Thus, by the strict definition of a trigonometric function (the hyper-geometric one), these are the simple trigonometric functions of this algebra. We call these functions the 3-dimensional H-type[88] simple-trig functions, and we denote them:

---

[88] The H is taken from hyperbolic with an eye on the 2-dimensional hyperbolic trigonometric functions. We take the view that hyperbolic space is a 2-dimensional

$$AH_3(x) = 1 + \frac{x^3}{3!} + \frac{x^6}{6!} + \ldots$$

$$BH_3(x) = x + \frac{x^4}{4!} + \frac{x^7}{7!} + \ldots \qquad (7.4)$$

$$CH_3(x) = \frac{x^2}{2!} + \frac{x^5}{5!} + \frac{x^8}{8!} + \ldots$$

Leading to:

$$\exp\left(\begin{bmatrix} a & b & c \\ c & a & b \\ b & c & a \end{bmatrix}\right) = PROD \left\{ \begin{bmatrix} r & 0 & 0 \\ 0 & r & 0 \\ 0 & 0 & r \end{bmatrix} \begin{bmatrix} AH_3(b) & BH_3(b) & CH_3(b) \\ CH_3(b) & AH_3(b) & BH_3(b) \\ BH_3(b) & CH_3(b) & AH_3(b) \end{bmatrix} \begin{bmatrix} AH_3(c) & CH_3(c) & BH_3(c) \\ BH_3(c) & AH_3(c) & CH_3(c) \\ CH_3(c) & BH_3(c) & AH_3(c) \end{bmatrix} \right\}$$

(7.5)

The matrices containing the H-type simple-trig functions were formed by exponentiating a matrix with $Trace = 0$. Since:

$$\det(\exp([\ ])) = \exp(Trace([\ ])) \qquad (7.6)$$

the determinants of these matrices are unity[89]. This leads, in both cases, to the same trigonometric identity analogous to the well-known $\cosh^2 x - \sinh^2 x = 1$. In the 3-dimensional case, this identity is:

Trigonometric Identity 1:
$$AH_3^{\ 3}(x) + BH_3^{\ 3}(x) + CH_3^{\ 3}(x) - 3AH_3(x)BH_3(x)CH_3(x) = 1 \quad (7.7)$$

We have:

---

thing and that we cannot apply that adjective to any spaces except the 2-dimensional one – hence the H.

[89] We will shortly give the sums of these series. With those sums, the fact that the determinant is unity can be shown by direct calculation.

Section 7

$$\begin{bmatrix} AH_3(b+x) & BH_3(b+x) & CH_3(b+x) \\ CH_3(b+x) & AH_3(b+x) & BH_3(b+x) \\ BH_3(b+x) & CH_3(b+x) & AH_3(b+x) \end{bmatrix} \quad (7.8)$$

$$= \exp\left(\begin{bmatrix} 0 & b+x & 0 \\ 0 & 0 & b+x \\ b+x & 0 & 0 \end{bmatrix}\right) = \exp\left(\begin{bmatrix} 0 & b & 0 \\ 0 & 0 & b \\ b & 0 & 0 \end{bmatrix} + \begin{bmatrix} 0 & x & 0 \\ 0 & 0 & x \\ x & 0 & 0 \end{bmatrix}\right)$$

$$= \exp\left(\begin{bmatrix} 0 & b & 0 \\ 0 & 0 & b \\ b & 0 & 0 \end{bmatrix}\right) \exp\left(\begin{bmatrix} 0 & x & 0 \\ 0 & 0 & x \\ x & 0 & 0 \end{bmatrix}\right)$$

$$= PROD \begin{bmatrix} AH_3(b) & BH_3(b) & CH_3(b) \\ CH_3(b) & AH_3(b) & BH_3(b) \\ BH_3(b) & CH_3(b) & AH_3(b) \end{bmatrix} \begin{bmatrix} AH_3(x) & BH_3(x) & CH_3(x) \\ CH_3(x) & AH_3(x) & BH_3(x) \\ BH_3(x) & CH_3(x) & AH_3(x) \end{bmatrix} \quad (7.9)$$

Thus, this matrix is a rotation matrix. By a similar procedure, the other matrix containing the H-type simple-trig functions is also a rotation matrix.

The reader has seen this procedure in the case of a 2-dimensional matrix in the chapter defining a rotation matrix. Clearly, it will work in all cases where a matrix is formed by exponentiation. In all cases, it will lead to a trigonometric identity, or three. In the cases of the two matrices considered here, we have:

Trigonometric Identity 2, 3, & 4:

$$AH_3(b+x) = AH_3(b)AH_3(x) + BH_3(b)CH_3(x) + CH_3(b)BH_3(x)$$
$$BH_3(b+x) = AH_3(b)BH_3(x) + BH_3(b)AH_3(x) + CH_3(b)CH_3(x) \quad (7.10)$$
$$CH_3(b+x) = AH_3(b)CH_3(x) + BH_3(b)BH_3(x) + CH_3(b)AH_3(x)$$

The exponentiated matrix is part of the algebra, and so we have:

$$\begin{bmatrix} x & y & z \\ z & x & y \\ y & z & x \end{bmatrix} = PROD \left\{ \begin{aligned} &\begin{bmatrix} r & 0 & 0 \\ 0 & r & 0 \\ 0 & 0 & r \end{bmatrix} \begin{bmatrix} AH_3(b) & BH_3(b) & CH_3(b) \\ CH_3(b) & AH_3(b) & BH_3(b) \\ BH_3(b) & CH_3(b) & AH_3(b) \end{bmatrix} \\ &\begin{bmatrix} AH_3(c) & CH_3(c) & BH_3(c) \\ BH_3(c) & AH_3(c) & CH_3(c) \\ CH_3(c) & BH_3(c) & AH_3(c) \end{bmatrix} \end{aligned} \right\} \quad (7.11)$$

$$\det\left(\begin{bmatrix} x & y & z \\ z & x & y \\ y & z & x \end{bmatrix}\right) = \det\left(\begin{bmatrix} r & 0 & 0 \\ 0 & r & 0 \\ 0 & 0 & r \end{bmatrix} \begin{bmatrix} AH_3(b) & BH_3(b) & CH_3(b) \\ CH_3(b) & AH_3(b) & BH_3(b) \\ BH_3(b) & CH_3(b) & AH_3(b) \end{bmatrix} \begin{bmatrix} AH_3(c) & CH_3(c) & BH_3(c) \\ BH_3(c) & AH_3(c) & CH_3(c) \\ CH_3(c) & BH_3(c) & AH_3(c) \end{bmatrix} \right)$$
(7.12)

Which is:

$$x^3 + y^3 + z^3 - 3xyz = r^3 \cdot 1 \cdot 1 \qquad (7.13)$$

Leading to the distance function:

$$r = \sqrt[3]{x^3 + y^3 + z^3 - 3xyz} \qquad (7.14)$$

This is the distance function forced upon us by the algebra. By the procedure above, it declares itself to be a distance function. This is a function of three independent real variables.

Note: Since $e^a > 0$ for all values of a, then $r^3 > 0$, and so $x^3 + y^3 + z^3 - 3xyz > 0$.
Hence, non-singularity is guaranteed.

The reader might expect that taking the cube root of a real number leads to one real root and two complex roots. However, this is not the 2-dimensional complex number algebra.

The H-type simple-trig functions appear in the rotation matrix. Consideration of the series representations of the H-type simple-trig functions shows that they have a differentiation cycle (of length three).

Section 7

We have a distance function, a set of two rotation matrices, and a set of three simple trigonometric functions. Multiplication of the two rotation matrices will give the compound trigonometric functions. We have a space.

We note that this algebra is comprised of the set of matrices that equate to the polar form. The determinant of the polar form can never be zero unless we have the zero matrix. Thus, this set of matrices is guaranteed non-singular, and we have no need of restrictions within the polar form. The restrictions in the Cartesian form arise because the space of this algebra does not cover 'all space'. This is analogous to the way 2-dimensional hyperbolic space does not fill the 2-dimensional Euclidean plane. In spite of not covering 'all space', this space, like the 2-dimensional hyperbolic space, is a complete and closed space.

The simple-trig functions have representation as standard functions.

$$AH_3(x) = \frac{1}{3}\left(e^x + 2e^{-\left(\frac{x}{2}\right)}\cos\left(\frac{\sqrt{3}}{2}x\right)\right)$$

$$BH_3(x) = \frac{1}{3}\left(e^x + e^{-\left(\frac{x}{2}\right)}\left(\sqrt{3}\sin\left(\frac{\sqrt{3}}{2}x\right) - \cos\left(\frac{\sqrt{3}}{2}x\right)\right)\right) \quad (7.15)$$

$$CH_3(x) = \frac{1}{3}\left(e^x - e^{-\left(\frac{x}{2}\right)}\left(\sqrt{3}\sin\left(\frac{\sqrt{3}}{2}x\right) + \cos\left(\frac{\sqrt{3}}{2}x\right)\right)\right)$$

The Polar Form of the $C_3 L^1 E^2_{(j=1,k=-1)}$ Algebra:

The polar form of the $C_3 L^1 E^2_{(j=1,k=-1)}$ algebra is:

$$\exp\left(\begin{bmatrix} a & b & c \\ c & a & -b \\ b & -c & a \end{bmatrix}\right) = \quad (7.16)$$

$$\text{PROD}\left(\begin{bmatrix} e^a & 0 & 0 \\ 0 & e^a & 0 \\ 0 & 0 & e^a \end{bmatrix}\begin{bmatrix} 1-\dfrac{b^3}{3!}+\dfrac{b^6}{6!}-\ldots & \dfrac{b}{1!}-\dfrac{b^4}{4!}+\dfrac{b^7}{7!}-\ldots & -\left(\dfrac{b^2}{2!}-\dfrac{b^5}{5!}+\dfrac{b^8}{8!}-\ldots\right) \\ -\left(\dfrac{b^2}{2!}-\dfrac{b^5}{5!}+\dfrac{b^8}{8!}-\ldots\right) & 1-\dfrac{b^3}{3!}+\dfrac{b^6}{6!}-\ldots & \dfrac{b}{1!}-\dfrac{b^4}{4!}+\dfrac{b^7}{7!}-\ldots \\ \dfrac{b}{1!}-\dfrac{b^4}{4!}+\dfrac{b^7}{7!}-\ldots & -\left(\dfrac{b^2}{2!}-\dfrac{b^5}{5!}+\dfrac{b^8}{8!}-\ldots\right) & 1-\dfrac{b^3}{3!}+\dfrac{b^6}{6!}-\ldots \end{bmatrix}\begin{bmatrix} 1-\dfrac{c^3}{3!}+\dfrac{c^6}{6!}-\ldots & -\left(\dfrac{c^2}{2!}-\dfrac{c^5}{5!}+\dfrac{c^8}{8!}-\ldots\right) & \dfrac{c}{1!}-\dfrac{c^4}{4!}+\dfrac{c^7}{7!}-\ldots \\ \dfrac{c}{1!}-\dfrac{c^4}{4!}+\dfrac{c^7}{7!}-\ldots & 1-\dfrac{c^3}{3!}+\dfrac{c^6}{6!}-\ldots & -\left(\dfrac{c^2}{2!}-\dfrac{c^5}{5!}+\dfrac{c^8}{8!}-\ldots\right) \\ -\left(\dfrac{c^2}{2!}-\dfrac{c^5}{5!}+\dfrac{c^8}{8!}-\ldots\right) & -\left(\dfrac{c}{1!}-\dfrac{c^4}{4!}+\dfrac{c^7}{7!}-\ldots\right) & 1-\dfrac{c^3}{3!}+\dfrac{c^6}{6!}-\ldots \end{bmatrix}\right)$$

(7.17)

The infinite series within the second two matrices are 3-way splittings of the exponential series with some minus signs chucked in. Thus, by the strict definition of a trigonometric function (the hyper-geometric one), these are simple trigonometric functions. We call these functions the 3-dimensional [90]E-type simple-trig functions, and we denote them:

$$AE_3(x) = 1 - \frac{x^3}{3!} + \frac{x^6}{6!} - \ldots$$

$$BE_3(x) = x - \frac{x^4}{4!} + \frac{x^7}{7!} - \ldots \quad (7.18)$$

$$CE_3(x) = \frac{x^2}{2!} - \frac{x^5}{5!} + \frac{x^8}{8!} - \ldots$$

Leading to:

---

[90] The E is taken from Euclidean with an eye on the 2-dimensional Euclidean trigonometric functions. We take the view that Euclidean space is a 2-dimensional thing and that we cannot apply that adjective to any spaces except the 2-dimensional one – hence the E.

Section 7

$$\exp\left(\begin{bmatrix} a & b & c \\ c & a & -b \\ b & -c & a \end{bmatrix}\right) = PROD \left\{ \begin{bmatrix} r & 0 & 0 \\ 0 & r & 0 \\ 0 & 0 & r \end{bmatrix} \begin{bmatrix} AE_3(b) & BE_3(b) & -CE_3(b) \\ -CE_3(b) & AE_3(b) & -BE_3(b) \\ BE_3(b) & CE_3(b) & AE_3(b) \end{bmatrix} \right\}$$
$$\begin{bmatrix} AE_3(c) & -CE_3(c) & BE_3(c) \\ BE_3(c) & AE_3(c) & CE_3(c) \\ -CE_3(c) & -BE_3(c) & AE_3(c) \end{bmatrix}$$
(7.19)

Again, the matrices containing the E-type simple-trig functions were formed by exponentiating a matrix with $Trace = 0$, and the determinants of these matrices are unity[91]. Both matrices lead to the following trigonometric identity.

Trigonometric Identity 1:
$$AE_3^{\ 3}(x) - BE_3^{\ 3}(x) + CE_3^{\ 3}(x) + 3AE_3(x)BE_3(x)CE_3(x) = 1 \quad (7.20)$$

We have:

$$\begin{bmatrix} AE_3(b+x) & BE_3(b+x) & -CE_3(b+x) \\ -CE_3(b+x) & AE_3(b+x) & -BE_3(b+x) \\ BE_3(b+x) & CE_3(b+x) & AE_3(b+x) \end{bmatrix}$$

$$= \exp\left(\begin{bmatrix} 0 & b+x & 0 \\ 0 & 0 & -(b+x) \\ b+x & 0 & 0 \end{bmatrix}\right) \quad (7.21)$$

$$= \exp\left(\begin{bmatrix} 0 & b & 0 \\ 0 & 0 & -b \\ b & 0 & 0 \end{bmatrix}\right) \exp\left(\begin{bmatrix} 0 & x & 0 \\ 0 & 0 & -x \\ x & 0 & 0 \end{bmatrix}\right)$$

$$= \begin{bmatrix} AE_3(b) & BE_3(b) & -CE_3(b) \\ -CE_3(b) & AE_3(b) & -BE_3(b) \\ BE_3(b) & CE_3(b) & AE_3(b) \end{bmatrix} \begin{bmatrix} AE_3(x) & BE_3(x) & -CE_3(x) \\ -CE_3(x) & AE_3(x) & -BE_3(x) \\ BE_3(x) & CE_3(x) & AE_3(x) \end{bmatrix}$$
(7.22)

---

[91] We will shortly give the sums of these series. With those sums, the fact that the determinant is unity can be shown by direct calculation.

Thus, this matrix is a rotation matrix. By a similar procedure, the other matrix containing the E-type simple-trig functions is also a rotation matrix. Both procedures lead to the same trigonometric identities.

Trigonometric Identity 2, 3, & 4:
$$AE_3(b+x) = AE_3(b)AE_3(x) - BE_3(b)CE_3(x) - CE_3(b)BE_3(x)$$
$$BE_3(b+x) = AE_3(b)BE_3(x) + BE_3(b)AE_3(x) - CE_3(b)CE_3(x) \quad (7.23)$$
$$CE_3(b+x) = AE_3(b)CE_3(x) + BE_3(b)BE_3(x) + CE_3(b)AE_3(x)$$

The distance function is:
$$r = \sqrt[3]{a^3 - b^3 - c^3 - 3abc} \quad (7.24)$$

Note:

Since $e^a > 0$ for all values of a, then $r^3 > 0$ and so $x^3 - y^3 - z^3 - 3xyz > 0$.

By the procedure above, this function declares itself to be a distance function.

The E-type simple-trig functions appear in the rotation matrix. Consideration of the series representations of the E-type simple-trig functions shows that they have a differentiation cycle (of length six).

We have a distance function, a set of two rotation matrices, and a set of three trigonometric functions. We have a natural space.

We note that this algebra is comprised of the set of matrices that equate to the polar form. The determinant of the polar form can never be zero unless we have the zero matrix. Thus, again, this set of matrices is guaranteed non-singular, and we have no need of restrictions within the polar form.

The simple-trig functions have representation as standard functions.

$$AE_3(x) = \frac{1}{3}\left(e^{-x} + 2e^{\left(\frac{x}{2}\right)}\cos\left(\frac{\sqrt{3}}{2}x\right)\right)$$

$$BE_3(x) = \frac{1}{3}\left(-e^{-x} + e^{\left(\frac{x}{2}\right)}\left(\sqrt{3}\sin\left(\frac{\sqrt{3}}{2}x\right) + \cos\left(\frac{\sqrt{3}}{2}x\right)\right)\right) \quad (7.25)$$

$$CE_3(x) = \frac{1}{3}\left(e^{-x} + e^{\left(\frac{x}{2}\right)}\left(\sqrt{3}\sin\left(\frac{\sqrt{3}}{2}x\right) - \cos\left(\frac{\sqrt{3}}{2}x\right)\right)\right)$$

Section 7

Polar Forms $C_3 L'E'H'_{(j=-1,k=1)}$ and $C_3 L'E'H'_{(j=-1,k=-1)}$ Algebras:

We have: $C_3 L'E'H'_{(j=-1,k=1)}$

$$\exp\left(\begin{bmatrix} a & b & c \\ -c & a & b \\ -b & -c & a \end{bmatrix}\right) = PROD \begin{Bmatrix} \begin{bmatrix} h & 0 & 0 \\ 0 & h & 0 \\ 0 & 0 & h \end{bmatrix} \begin{bmatrix} AE_3(b) & BE_3(b) & CE_3(b) \\ -CE_3(b) & AE_3(b) & BE_3(b) \\ -BE_3(b) & -CE_3(b) & AE_3(b) \end{bmatrix} \\ \begin{bmatrix} AH_3(c) & -CH_3(c) & BH_3(c) \\ -BH_3(c) & AH_3(c) & -CH_3(c) \\ CH_3(c) & -BH_3(c) & AH_3(c) \end{bmatrix} \end{Bmatrix}$$

(7.26)

and: $C_3 L'E'H'_{(j=-1,k=-1)}$:

$$\exp\left(\begin{bmatrix} a & b & c \\ -c & a & -b \\ -b & c & a \end{bmatrix}\right) = PROD \begin{Bmatrix} \begin{bmatrix} h & 0 & 0 \\ 0 & h & 0 \\ 0 & 0 & h \end{bmatrix} \begin{bmatrix} AH_3(b) & BH_3(b) & -CH_3(b) \\ CH_3(b) & AH_3(b) & -BH_3(b) \\ -BH_3(b) & -CH_3(b) & AH_3(b) \end{bmatrix} \\ \begin{bmatrix} AE_3(c) & CE_3(c) & BE_3(c) \\ -BE_3(c) & AE_3(c) & -CE_3(c) \\ -CE_3(c) & BE_3(c) & AE_3(c) \end{bmatrix} \end{Bmatrix}$$

(7.27)

The determinants of the rotation matrices being unity leads to the trigonometric identities previously found. The distance functions are, respectively:

$$C_3 L'E'H'_{(j=-1,k=1)} \quad : \quad r = \sqrt[3]{a^3 - b^3 + c^3 + 3abc}$$

$$C_3 L'E'H'_{(j=-1,k=-1)} \quad : \quad r = \sqrt[3]{a^3 + b^3 - c^3 + 3abc}$$

(7.28)

Again, in each case, we have the requirements of a space, but these spaces are mixed spaces containing both H-type angles and E-Type angles.

The 3-dimensional v-functions[92] - an Introduction:

The reader will doubtless have noticed that the 3-dimensional algebras have two simple rotation matrices in the polar form. What do we get if these two matrices are multiplied together?

---

[92] Pronounced the nu-functions. So named because they were unknown before they were discovered. And a terrible jest it is to readers who are not fluent in English.

We have: $C_3 L'H^2_{(j=1,k=1)}$:

$$\exp\left(\begin{bmatrix} a & b & c \\ c & a & b \\ b & c & a \end{bmatrix}\right) = \text{PROD}\left\{\begin{bmatrix} r & 0 & 0 \\ 0 & r & 0 \\ 0 & 0 & r \end{bmatrix}\begin{bmatrix} AH_3(b) & BH_3(b) & CH_3(b) \\ CH_3(b) & AH_3(b) & BH_3(b) \\ BH_3(b) & CH_3(b) & AH_3(b) \end{bmatrix}\begin{bmatrix} AH_3(c) & CH_3(c) & BH_3(c) \\ BH_3(c) & AH_3(c) & CH_3(c) \\ CH_3(c) & BH_3(c) & AH_3(c) \end{bmatrix}\right\}$$

(7.29)

We define:

$$v_{[C_3 L'H'_{(j=1,k=1)}]} A(b,c) = AH_3(b)AH_3(c) + BH_3(b)BH_3(c) + CH_3(b)CH_3(c)$$

$$v_{[C_3 L'H'_{(j=1,k=1)}]} B(b,c) = AH_3(c)BH_3(b) + AH_3(b)CH_3(c) + BH_3(c)CH_3(b)$$

$$v_{[C_3 L'H'_{(j=1,k=1)}]} C(b,c) = AH_3(c)CH_3(b) + AH_3(b)BH_3(c) + BH_3(b)CH_3(c)$$

(7.30)

The subscripted bracket contains the name of the algebra, and A, B, C designate the different functions. The A, B, C correspond to the position in the Cartesian algebraic matrix form. With this notation, the compound polar form is:

$$\exp\left(\begin{bmatrix} a & b & c \\ c & a & b \\ b & c & a \end{bmatrix}\right) = \text{PROD}\left\{\begin{bmatrix} r & 0 & 0 \\ 0 & r & 0 \\ 0 & 0 & r \end{bmatrix}\begin{bmatrix} v_{[C_3 i H'_{(j=1,k=1)}]}A(b,c) & v_{[C_3 i H'_{(j=1,k=1)}]}B(b,c) & v_{[C_3 i H'_{(j=1,k=1)}]}C(b,c) \\ v_{[C_3 i H'_{(j=1,k=1)}]}C(b,c) & v_{[C_3 i H'_{(j=1,k=1)}]}A(b,c) & v_{[C_3 i H'_{(j=1,k=1)}]}B(b,c) \\ v_{[C_3 i H'_{(j=1,k=1)}]}B(b,c) & v_{[C_3 i H'_{(j=1,k=1)}]}C(b,c) & v_{[C_3 i H'_{(j=1,k=1)}]}A(b,c) \end{bmatrix}\right\}$$

(7.31)

Clearly, the determinant of this matrix, being the product of two rotation matrices, is unity.

---

v-function Identity 1:

Section 7

$$\left(v_{[c,i,H'_{(r+s,s)}]}A(b,c)\right)^3 + \left(v_{[c,i,H'_{(r+s,s)}]}B(b,c)\right)^3 + \left(v_{[c,i,H'_{(r+s,s)}]}C(b,c)\right)^3 \\ -3v_{[c,i,H'_{(r+s,s)}]}A(b,c)v_{[c,i,H'_{(r+s,s)}]}B(b,c)v_{[c,i,H'_{(r+s,s)}]}C(b,c) = 1 \quad (7.32)$$

We have:

$$\begin{bmatrix} v_{[c,i,H'_{(r+s,s)}]}A(b+e,c+f) & v_{[c,i,H'_{(r+s,s)}]}B(b+e,c+f) & v_{[c,i,H'_{(r+s,s)}]}C(b+e,c+f) \\ v_{[c,i,H'_{(r+s,s)}]}C(b+e,c+f) & v_{[c,i,H'_{(r+s,s)}]}A(b+e,c+f) & v_{[c,i,H'_{(r+s,s)}]}B(b+e,c+f) \\ v_{[c,i,H'_{(r+s,s)}]}B(b+e,c+f) & v_{[c,i,H'_{(r+s,s)}]}C(b+e,c+f) & v_{[c,i,H'_{(r+s,s)}]}A(b+e,c+f) \end{bmatrix}$$

$$= \exp\left(\begin{bmatrix} 0 & b+e & c+f \\ c+f & 0 & b+e \\ b+e & c+f & 0 \end{bmatrix}\right)$$

$$= \exp\left(\begin{bmatrix} 0 & b & c \\ c & 0 & b \\ b & c & 0 \end{bmatrix}\right)\exp\left(\begin{bmatrix} 0 & e & f \\ f & 0 & e \\ e & f & 0 \end{bmatrix}\right)$$

(7.33)

$$= PROD\left(\begin{bmatrix} v_{[c,i,H'_{(r+s,s)}]}A(b,c) & v_{[c,i,H'_{(r+s,s)}]}B(b,c) & v_{[c,i,H'_{(r+s,s)}]}C(b,c) \\ v_{[c,i,H'_{(r+s,s)}]}C(b,c) & v_{[c,i,H'_{(r+s,s)}]}A(b,c) & v_{[c,i,H'_{(r+s,s)}]}B(b,c) \\ v_{[c,i,H'_{(r+s,s)}]}B(b,c) & v_{[c,i,H'_{(r+s,s)}]}C(b,c) & v_{[c,i,H'_{(r+s,s)}]}A(b,c) \end{bmatrix}\right. \\ \left.\begin{bmatrix} v_{[c,i,H'_{(r+s,s)}]}A(e,f) & v_{[c,i,H'_{(r+s,s)}]}B(e,f) & v_{[c,i,H'_{(r+s,s)}]}C(e,f) \\ v_{[c,i,H'_{(r+s,s)}]}C(e,f) & v_{[c,i,H'_{(r+s,s)}]}A(e,f) & v_{[c,i,H'_{(r+s,s)}]}B(e,f) \\ v_{[c,i,H'_{(r+s,s)}]}B(e,f) & v_{[c,i,H'_{(r+s,s)}]}C(e,f) & v_{[c,i,H'_{(r+s,s)}]}A(e,f) \end{bmatrix}\right)$$

(7.34)

Leading to:

v-function Identity 2, 3, & 4:

$$v_{\left[C_3 l'H^2_{(J-1,k-1)}\right]}A(b+e,c+f) = v_{[\ ]}A(b,c)v_{[\ ]}A(e,f)$$
$$+v_{[\ ]}B(b,c)v_{[\ ]}C(e,f)+v_{[\ ]}C(b,c)v_{[\ ]}B(e,f)$$

$$v_{\left[C_3 l'H^2_{(J-1,k-1)}\right]}B(b+e,c+f) = v_{[\ ]}B(b,c)v_{[\ ]}A(e,f)$$
$$+v_{[\ ]}C(b,c)v_{[\ ]}C(e,f)+v_{[L]}A(b,c)v_{[\ ]}B(e,f)$$

$$v_{\left[C_3 l'H^2_{(J-1,k-1)}\right]}C(b+e,c+f) = v_{[\ ]}C(b,c)v_{[\ ]}A(e,f)$$
$$+v_{[\ ]}A(b,c)v_{[\ ]}C(e,f)+v_{[\ ]}B(b,c)v_{[\ ]}B(e,f)$$

(7.35)

Thus the compound matrix has the properties necessary to be a rotation matrix.

In standard form:

$$v_{\left[c_3 l' H'_{(\ldots)}\right]}A(b,c) = \frac{1}{3}\left(e^{(b+c)} + 2e^{-\left(\frac{b+c}{2}\right)}\cos\left(\frac{\sqrt{3}}{2}(b-c)\right)\right) \quad (7.36)$$

$$v_{\left[c_3 l' H'_{(\ldots)}\right]}B(b,c) = \frac{1}{3}\left(e^{(b+c)} + e^{-\left(\frac{b+c}{2}\right)}\left(\sqrt{3}\sin\left(\frac{\sqrt{3}}{2}(b-c)\right) - \cos\left(\frac{\sqrt{3}}{2}(b-c)\right)\right)\right)$$
(7.37)

$$v_{\left[c_3 l' H'_{(\ldots)}\right]}C(b,c) = \frac{1}{3}\left(e^{(b+c)} - e^{-\left(\frac{b+c}{2}\right)}\left(\sqrt{3}\sin\left(\frac{\sqrt{3}}{2}(b-c)\right) + \cos\left(\frac{\sqrt{3}}{2}(b-c)\right)\right)\right)$$
(7.38)

The reader should compare these to the standard forms of the 3-dimensional simple-trig functions. These are not the simple-trig functions with a double argument. There is no misprinting of the signs of $\{(b+c),(b-c)\}$.

We point out that:

$$v_{\left[c_3 l' H'_{(\ldots)}\right]}A(b,c) + v_{\left[c_3 l' H'_{(\ldots)}\right]}B(b,c) + v_{\left[lc_3 l' H'_{(\ldots)}\right]}C(b,c) = e^{(b+c)} \quad (7.39)$$

and so the v-functions are, at least in this sense, splittings of the exponential.

We have:

Section 7

$$\begin{bmatrix} x & y & z \\ z & x & y \\ y & z & x \end{bmatrix} = \begin{bmatrix} r & 0 & 0 \\ 0 & r & 0 \\ 0 & 0 & r \end{bmatrix} \begin{bmatrix} v_{[C_iL H^i_{(j,i,k)}]} A(b,c) & v_{[C_iL H^i_{(j,i,k)}]} B(b,c) & v_{[C_iL H^i_{(j,i,k)}]} C(b,c) \\ v_{[C_iL H^i_{(j,i,k)}]} C(b,c) & v_{[C_iL H^i_{(j,i,k)}]} A(b,c) & v_{[C_iL H^i_{(j,i,k)}]} B(b,c) \\ v_{[C_iL H^i_{(j,i,k)}]} B(b,c) & v_{[C_iL H^i_{(j,i,k)}]} C(b,c) & v_{[C_iL H^i_{(j,i,k)}]} A(b,c) \end{bmatrix} \quad (7.40)$$

Normalising this brings us on to the unit sphere:

$$\frac{1}{\sqrt[3]{x^3+y^3+z^3-3xyz}} \begin{bmatrix} x & y & z \\ z & x & y \\ y & z & x \end{bmatrix} = \quad (7.41)$$

$$\begin{bmatrix} v_{[C_iL H^i_{(j,i,k)}]} A(b,c) & v_{[C_iL H^i_{(j,i,k)}]} B(b,c) & v_{[C_iL H^i_{(j,i,k)}]} C(b,c) \\ v_{[C_iL H^i_{(j,i,k)}]} C(b,c) & v_{[C_iL H^i_{(j,i,k)}]} A(b,c) & v_{[C_iL H^i_{(j,i,k)}]} B(b,c) \\ v_{[C_iL H^i_{(j,i,k)}]} B(b,c) & v_{[C_iL H^i_{(j,i,k)}]} C(b,c) & v_{[C_iL H^i_{(j,i,k)}]} A(b,c) \end{bmatrix} \quad (7.42)$$

Leading to:

$$\frac{x}{\sqrt[3]{x^3+y^3+z^3-3xyz}} = v_{[\ ]} A$$

$$\frac{y}{\sqrt[3]{x^3+y^3+z^3-3xyz}} = v_{[\ ]} B \quad (7.43)$$

$$\frac{z}{\sqrt[3]{x^3+y^3+z^3-3xyz}} = v_{[\ ]} C$$

And so, these are functions are projection functions from the unit sphere on to the axes. By similar means, we can define v-functions for the other algebras.

The $C_3 L^1 E^2_{(j=1,k=-1)}$ algebra:

$$v_{[C_3 L^I E^I_{(j=1,k=1)}]} A(b,c) = \frac{1}{3}\left(e^{(b+c)} + 2e^{\left(\frac{b+c}{2}\right)}\cos\left(\frac{\sqrt{3}}{2}(b-c)\right)\right)$$

$$v_{[C_3 L^I E^I_{(j=1,k=1)}]} B(b,c) = \frac{1}{3}\left(-e^{(b+c)} + e^{\left(\frac{b+c}{2}\right)}\left(\sqrt{3}\sin\left(\frac{\sqrt{3}}{2}(b-c)\right) + \cos\left(\frac{\sqrt{3}}{2}(b-c)\right)\right)\right)$$

$$v_{[C_3 L^I E^I_{(j=1,k=1)}]} C(b,c) = \frac{1}{3}\left(-e^{(b+c)} - e^{\left(\frac{b+c}{2}\right)}\left(\sqrt{3}\sin\left(\frac{\sqrt{3}}{2}(b-c)\right) - \cos\left(\frac{\sqrt{3}}{2}(b-c)\right)\right)\right)$$

(7.44)

The $C_3 L^I E^I H^I_{(J=-1,k=1)}$ algebra:

$$v_{[C_3 L^I E^I H^I_{(j=-1,k=1)}]} A(b,c) = \frac{1}{3}\left(e^{(b-c)} + 2e^{\left(\frac{b-c}{2}\right)}\cos\left(\frac{\sqrt{3}}{2}(b+c)\right)\right)$$

$$v_{[C_3 L^I E^I H^I_{(j=-1,k=1)}]} B(b,c) = \frac{1}{3}\left(-e^{(b-c)} + e^{\left(\frac{b-c}{2}\right)}\left(\sqrt{3}\sin\left(\frac{\sqrt{3}}{2}(b+c)\right) + \cos\left(\frac{\sqrt{3}}{2}(b+c)\right)\right)\right)$$

$$v_{[C_3 L^I E^I H^I_{(j=-1,k=1)}]} C(b,c) = \frac{1}{3}\left(e^{(b-c)} + e^{\left(\frac{b-c}{2}\right)}\left(\sqrt{3}\sin\left(\frac{\sqrt{3}}{2}(b+c)\right) - \cos\left(\frac{\sqrt{3}}{2}(b+c)\right)\right)\right)$$

(7.45)

The $C_3 L^I E^I H^I_{(J=-1,k=-1)}$ algebra:

$$v_{[C_3 L^I E^I H^I_{(j=-1,k=-1)}]} A(b,c) = \frac{1}{3}\left(e^{(b-c)} + 2e^{\left(\frac{b-c}{2}\right)}\cos\left(\frac{\sqrt{3}}{2}(b+c)\right)\right)$$

$$v_{[C_3 L^I E^I H^I_{(j=-1,k=-1)}]} B(b,c) = \frac{1}{3}\left(e^{(b-c)} + e^{\left(\frac{b-c}{2}\right)}\left(\sqrt{3}\sin\left(\frac{\sqrt{3}}{2}(b+c)\right) - \cos\left(\frac{\sqrt{3}}{2}(b+c)\right)\right)\right)$$

$$v_{[C_3 L^I E^I H^I_{(j=-1,k=-1)}]} C(b,c) = \frac{1}{3}\left(-e^{(b-c)} + e^{\left(\frac{b-c}{2}\right)}\left(\sqrt{3}\sin\left(\frac{\sqrt{3}}{2}(b+c)\right) + \cos\left(\frac{\sqrt{3}}{2}(b+c)\right)\right)\right)$$

(7.46)

The v-functions will be investigated more thoroughly in later chapters.

The General Polar Form:

The general form of the 3-dimensional algebraic matrix forms is:

Section 7

$$C_3 = \begin{bmatrix} a & b & c \\ jc & a & kb \\ jb & \dfrac{j}{k}c & a \end{bmatrix} : \{j, k \neq 0\} \qquad (7.47)$$

We have:

$$\exp\begin{bmatrix} a & b & c \\ jc & a & kb \\ jb & \dfrac{j}{k}c & a \end{bmatrix} = \exp\begin{bmatrix} a & 0 & 0 \\ 0 & a & 0 \\ 0 & 0 & a \end{bmatrix} \exp\begin{bmatrix} 0 & b & 0 \\ 0 & 0 & kb \\ jb & 0 & 0 \end{bmatrix} \exp\begin{bmatrix} 0 & 0 & c \\ jc & 0 & 0 \\ 0 & \dfrac{j}{k}c & 0 \end{bmatrix}$$

(7.48)

$$= \text{PROD} \left(\begin{bmatrix} e^a & 0 & 0 \\ 0 & e^a & 0 \\ 0 & 0 & e^a \end{bmatrix} \begin{bmatrix} 1 + jk\dfrac{b^3}{3!} + j^2k^2\dfrac{b^6}{6!} + \ldots & b + jk\dfrac{b^4}{4!} + \ldots & k\dfrac{b^2}{2!} + jk^2\dfrac{b^5}{5!} + \ldots \\ j\left(k\dfrac{b^2}{2!} + jk^2\dfrac{b^5}{5!} + \ldots\right) & 1 + jk\dfrac{b^3}{3!} + j^2k^2\dfrac{b^6}{6!} + \ldots & k\left(b + jk\dfrac{b^4}{4!} \ldots\right) \\ j\left(b + jk\dfrac{b^4}{4!} + \ldots\right) & \dfrac{j}{k}\left(k\dfrac{b^2}{2!} + jk^2\dfrac{b^5}{5!} + \ldots\right) & 1 + jk\dfrac{b^3}{3!} + \ldots \end{bmatrix} \begin{bmatrix} 1 + \dfrac{j\ c^3}{k\ 3!} + \dfrac{j^2\ c^6}{k^2\ 6!} + \ldots & \dfrac{j\ c^2}{k\ 2!} + \dfrac{j^2\ c^5}{k^2\ 5!} + \ldots & c + \dfrac{j\ c^4}{k\ 4!} + \dfrac{j^2\ c^7}{k^2\ 7!} + \ldots \\ j\left(c + \dfrac{j\ c^4}{k\ 4!} + \ldots\right) & 1 + \dfrac{j\ c^3}{k\ 3!} + \ldots & k\left(\dfrac{j\ c^2}{k\ 2!} + \dfrac{j^2\ c^5}{k^2\ 5!} + \ldots\right) \\ j\left(\dfrac{j\ c^2}{k\ 2!} + \dfrac{j^2\ c^5}{k^2\ 5!} + \ldots\right) & \dfrac{j}{k}\left(c + \dfrac{j\ c^4}{k\ 4!} + \ldots\right) & 1 + \dfrac{j\ c^3}{k\ 3!} + \ldots \end{bmatrix} \right)$$

(7.49)

The determinants of the rotation matrices, in both cases, are:

$$AH_3^3(\ ) + BH_3^3(\ ) + CH_3^3(\ ) - 3AH_3^3(\ )BH_3^3(\ )CH_3^3(\ ) = 1 \qquad (7.50)$$

The distance function is:

$$r = \sqrt[3]{a^3 + jkb^3 + \dfrac{j^2}{k}c^3 - 3jabc} \qquad (7.51)$$

When $\{j=1, k=-1\}$, this general form becomes the $C_3L'E^2_{(j=1,k=-1)}$ algebra. We equate the b rotation matrices of these two cases.

$$\begin{bmatrix} AH_3\left((-1)^{\frac{1}{3}}b\right) & \frac{1}{(-1)^{\frac{1}{3}}}BH_3\left((-1)^{\frac{1}{3}}b\right) & \frac{-1}{(-1)^{\frac{2}{3}}}CH_3\left((-1)^{\frac{1}{3}}b\right) \\ \frac{-1}{(-1)^{\frac{2}{3}}}CH_3\left((-1)^{\frac{1}{3}}b\right) & AH_3\left((-1)^{\frac{1}{3}}b\right) & -\left(\frac{1}{(-1)^{\frac{1}{3}}}BH_3\left((-1)^{\frac{1}{3}}b\right)\right) \\ \left(\frac{1}{(-1)^{\frac{1}{3}}}BH_3\left((-1)^{\frac{1}{3}}b\right)\right) & -\left(\frac{-1}{(-1)^{\frac{2}{3}}}CH_3\left((-1)^{\frac{1}{3}}b\right)\right) & AH_3\left((-1)^{\frac{1}{3}}b\right) \end{bmatrix}$$
(7.52)

$$= \begin{bmatrix} AE_3(b) & BE_3(b) & -CE_3(b) \\ -CE_3(b) & AE_3(b) & -BE_3(b) \\ BE_3(b) & CE_3(b) & AE_3(b) \end{bmatrix}$$
(7.53)

Leading to:

$$AH_3\left((-1)^{\frac{1}{3}}b\right) = AE_3(b)$$

$$BH_3\left((-1)^{\frac{1}{3}}b\right) = (-1)^{\frac{1}{3}} BE_3(b) \quad (7.54)$$

$$CH_3\left((-1)^{\frac{1}{3}}b\right) = (1)^{\frac{1}{3}} CE_3(b)$$

which is analogous to the 2-dimensional case relating the Euclidean trigonometric functions to the hyperbolic trigonometric functions. The same results can be derived from any of the other algebras previously discussed.

The C-R Equations for the $C_3L'H^2_{(j=1,k=1)}$ Algebra:

Consider the 3 by 3 matrix function:

$$C_3L'H^2_{(j=1,k=1)} : f\left(\begin{bmatrix} a & b & c \\ c & a & b \\ b & c & a \end{bmatrix}\right) \rightarrow \begin{bmatrix} u(a,b,c) & v(a,b,c) & w(a,b,c) \\ w(a,b,c) & u(a,b,c) & v(a,b,c) \\ v(a,b,c) & w(a,b,c) & u(a,b,c) \end{bmatrix}$$
(7.55)

Section 7

We differentiate with respect to $\begin{bmatrix} a & 0 & 0 \\ 0 & a & 0 \\ 0 & 0 & a \end{bmatrix}$ (holding b, c constant). Since this matrix is a real number, we have:

$$f'\left(\begin{bmatrix} a & b & c \\ c & a & b \\ b & c & a \end{bmatrix}\right) = \begin{bmatrix} \dfrac{\partial u}{\partial a} & \dfrac{\partial v}{\partial a} & \dfrac{\partial w}{\partial a} \\ \dfrac{\partial w}{\partial a} & \dfrac{\partial u}{\partial a} & \dfrac{\partial v}{\partial a} \\ \dfrac{\partial v}{\partial a} & \dfrac{\partial w}{\partial a} & \dfrac{\partial u}{\partial a} \end{bmatrix} \qquad (7.56)$$

We now differentiate with respect to $\begin{bmatrix} 0 & b & 0 \\ 0 & 0 & b \\ b & 0 & 0 \end{bmatrix} = \begin{bmatrix} 0 & 1 & 0 \\ 0 & 0 & 1 \\ 1 & 0 & 0 \end{bmatrix} \begin{bmatrix} b & 0 & 0 \\ 0 & b & 0 \\ 0 & 0 & b \end{bmatrix}$

We have:

$$f'\left(\begin{bmatrix} a & b & c \\ c & a & b \\ b & c & a \end{bmatrix}\right)_b = \begin{bmatrix} \dfrac{\partial v}{\partial b} & \dfrac{\partial w}{\partial b} & \dfrac{\partial u}{\partial b} \\ \dfrac{\partial u}{\partial b} & \dfrac{\partial v}{\partial b} & \dfrac{\partial w}{\partial b} \\ \dfrac{\partial w}{\partial b} & \dfrac{\partial u}{\partial b} & \dfrac{\partial v}{\partial b} \end{bmatrix} \qquad (7.57)$$

We now differentiate with respect to $\begin{bmatrix} 0 & 0 & c \\ c & 0 & 0 \\ 0 & c & 0 \end{bmatrix} = \begin{bmatrix} 0 & 0 & 1 \\ 1 & 0 & 0 \\ 0 & 1 & 0 \end{bmatrix} \begin{bmatrix} c & 0 & 0 \\ 0 & c & 0 \\ 0 & 0 & c \end{bmatrix}$

We have:

$$f'\left(\begin{bmatrix} a & b & c \\ c & a & b \\ b & c & a \end{bmatrix}\right)_c = \begin{bmatrix} \dfrac{\partial w}{\partial c} & \dfrac{\partial u}{\partial c} & \dfrac{\partial v}{\partial c} \\ \dfrac{\partial v}{\partial c} & \dfrac{\partial w}{\partial c} & \dfrac{\partial u}{\partial c} \\ \dfrac{\partial u}{\partial c} & \dfrac{\partial v}{\partial c} & \dfrac{\partial w}{\partial c} \end{bmatrix} \qquad (7.58)$$

Putting these together, a function, $f(\ )$, is differentiable only if these matrices are equal. Thus, the Cauchy Riemann equations for the $C_3 L^1 H^2_{(j=1,k=1)}$ algebra are:

$$\frac{\partial u}{\partial a} = \frac{\partial v}{\partial b} = \frac{\partial w}{\partial c}$$
$$\frac{\partial v}{\partial a} = \frac{\partial w}{\partial b} = \frac{\partial u}{\partial c} \quad (7.59)$$
$$\frac{\partial w}{\partial a} = \frac{\partial u}{\partial b} = \frac{\partial v}{\partial c}$$

Non-Matrix Notation of the Algebras:

> 2nd edition note: The non-matrix notation of the algebras has found no usefulness in later developments of this area of mathematics. Of course, one could never see into the future.

Although we are using matrix notation to do our algebras, it is sometimes useful[93] and time saving to revert to non-matrix notation. We find the non-matrix notation by inspection of the matrix notation. We demonstrate with the complex numbers and a 3-dimensional algebra.

Addition: In all the natural algebras, addition is component-wise.

Complex Numbers:
This is a 2-dimensional algebra. In non-matrix notation, we can denote one axis by the absence of any symbol but we need a symbol to denote the other axis. We choose a symbol with a hat on[94], $\hat{i}$.

Multiplication: We obtain the multiplication operation by inspection of the top row of the product matrix:

$$\begin{bmatrix} a & b \\ -b & a \end{bmatrix} \begin{bmatrix} c & d \\ -d & c \end{bmatrix} = \begin{bmatrix} ac-bd & ad+bc \\ \sim & \sim \end{bmatrix} \quad (7.60)$$

---

[93] Particularly in proofs of algebraic isomorphism
[94] As I write, the weather is fiercely cold. It would be unwise and cruel to choose a symbol that is not properly dressed.

Section 7

$$(a,\hat{i}b)(c,\hat{i}d) = ((ac-bd),\hat{i}(ad+bc))$$
$$(a+\hat{i}b)(c+\hat{i}d) = (ac-bd)+\hat{i}(ad+bc)$$
(7.61)

The reader will note that, in the third line of the above, we have used the accepted notational nonsense of inserting a + sign where there ought not to be one. Nonsense though it is, we will also use it elsewhere.

Conjugation: The conjugate of an algebraic matrix form is the adjoint matrix of that algebraic matrix form. The conjugate is obtained from the top row of the adjoint matrix:

$$adj\left(\begin{bmatrix} a & b \\ -b & a \end{bmatrix}\right) = \begin{bmatrix} a & -b \\ b & a \end{bmatrix}$$
(7.62)

$$\overline{(a,\hat{i}b)} = (a,-\hat{i}b)$$
$$\overline{(a+\hat{i}b)} = a-\hat{i}b$$
(7.63)

Since the adjoint matrix is the inverse matrix multiplied by the determinant, the product of the matrix and its conjugate is the determinant. We use this to discover the nature of the $\hat{i}$. To do this, we multiply the number and its conjugate together as if the variables were all real numbers; we do not use the multiplication operation taken from the matrix:

$$(a+\hat{i}b)(a-\hat{i}b) = a^2 - a\hat{i}b + \hat{i}ba - \hat{i}^2 b^2 = a^2 - \hat{i}^2 b^2$$
(7.64)

Equating this to the determinant of the algebraic matrix form, $a^2 + b^2$, leads to $\hat{i}^2 = -1$.[95]

The 3-dimensional Case:

These are 3-dimensional algebras. One of the dimensions is the real numbers; to denote the other two dimensions, we need two symbols with hats on, $\hat{p}, \hat{q}$. We use the same two symbols in each of the four flat algebras. Since these symbols mean different things in each of the different flat algebras, this is a potential trap for the unwary. However, we feel it is preferable to the alternative option of potentially confusing the wary.

Addition is component-wise.
$$(a+\hat{p}b+\hat{q}c)+(d+\hat{p}e+\hat{q}f) = (a+d)+\hat{p}(b+e)+\hat{q}(c+f)$$ (7.65)

---

[95] If we were clever, we could achieve the same by comparing the two products.

Multiplication:

$$\begin{bmatrix} a & b & c \\ jc & a & kb \\ jb & \left(\dfrac{j}{k}\right)c & a \end{bmatrix} \begin{bmatrix} d & e & f \\ jf & d & ke \\ je & \left(\dfrac{j}{k}\right)f & d \end{bmatrix} \quad (7.66)$$

$$= \begin{bmatrix} ad + jbf + jce & ae + bd + \left(\dfrac{j}{k}\right)cf & af + kbe + cd \\ & & \\ & & \end{bmatrix} \quad (7.67)$$

$$(a + \hat{p}b + \hat{q}c)(d + \hat{p}e + \hat{q}f)$$
$$= (ad + jbf + jce) + \hat{p}\left(ae + bd + \left(\dfrac{j}{k}\right)cf\right) + \hat{q}(af + kbe + cd) \quad (7.68)$$

Conjugation:

$$\mathrm{adj}\left(\begin{bmatrix} a & b & c \\ jc & a & kb \\ jb & \left(\dfrac{j}{k}\right)c & a \end{bmatrix}\right) = \begin{bmatrix} a^2 - jbc & \left(\dfrac{j}{k}\right)c^2 - ab & kb^2 - ac \\ & & \\ & & \end{bmatrix}$$

$$\overline{(a + pb + \hat{q}c)} = (a^2 - jbc) + \hat{p}\left(\left(\dfrac{j}{k}\right)c^2 - ab\right) + \hat{q}(kb^2 - ac)$$
$$(7.69)$$

The product of the 3-dimensional complex number and its conjugate is the determinant. We multiply the two together as if the variables were real numbers.

$$(a + pb + \hat{q}c)\left((a^2 - jbc) + p\left(\left(\dfrac{j}{k}\right)c^2 - ab\right) + \hat{q}(kb^2 - ac)\right)$$
$$= (a^3 - jabc) + p\left(\left(\dfrac{j}{k}\right)ac^2 - a^2b\right) + \hat{q}(kab^2 - a^2c)$$
$$+ p(a^2b - jb^2c) + p^2\left(\left(\dfrac{j}{k}\right)bc^2 - ab^2\right) + p\hat{q}(kb^3 - abc) \quad (7.70)$$
$$+ \hat{q}(a^2c - jbc^2) + \hat{q}p\left(\left(\dfrac{j}{k}\right)c^3 - abc\right) + \hat{q}^2(kb^2c - ac^2)$$

Leading to:

119

Section 7

$$a^3 + p\hat{q}kb^3 + \hat{q}p\left(\frac{j}{k}\right)c^3 + abc\left(-j - p\hat{q} - \hat{q}p\right) + p\left(\left(\frac{j}{k}\right)ac^2 - jb^2c\right)$$

$$+\hat{q}\left(kab^2 - jbc^2\right) + p^2\left(\left(\frac{j}{k}\right)bc^2 - ab^2\right) + \hat{q}^2\left(kb^2c - ac^2\right)$$

(7.71)

There are four different flat algebras depending upon the values of $\{j,k\}$, but all four flat algebras have determinants with the four terms $\{a^3, b^3, c^3, 3abc\}$; only the signs vary. We take them case-by-case and compare the above product to the determinant. We get:

Case: $j = k = +1$:

$$a^3 + p\hat{q}b^3 + \hat{q}pc^3 + abc\left(-1 - p\hat{q} - \hat{q}p\right)$$

$$+ p\left(ac^2 - b^2c\right) + \hat{q}\left(ab^2 - bc^2\right) \qquad (7.72)$$

$$+ p^2\left(bc^2 - ab^2\right) + \hat{q}^2\left(b^2c - ac^2\right)$$

Now:

$$\det\left(C_3 L^1 H^2_{(j=1,k=1)}\right) = a^3 + b^3 + c^3 - 3abc \qquad (7.73)$$

which compels us to conclude that:

$$\left\{\left\{\hat{q}p = p\hat{q} = +1, p^2 = \hat{q}, \hat{q}^2 = p\right\} \Rightarrow \left\{p^3 = +1, \hat{q}^3 = +1\right\}\right\} \qquad (7.74)$$

The other flat cases lead to:

$$C_3 L' E^2_{(j=1,k=-1)}$$

$$\left\{p\hat{q} = \hat{q}p = +1, \hat{q}^2 = -p, p^2 = -\hat{q}\right\} \qquad (7.75)$$

$$\left\{p^3 = -1, \hat{q}^3 = -1\right\}$$

$$C_3 L' E^1 H^1_{(j=-1,k=1)}$$

$$\left\{p\hat{q} = \hat{q}p = -1, \hat{q}^2 = -p, p^2 = +\hat{q}\right\} \qquad (7.76)$$

$$\left\{p^3 = -1, \hat{q}^3 = +1\right\}$$

$C_3 L^1 E^1 H^1{}_{(j=-1,k=-1)}$

$$\left\{ p\hat{q} = \hat{q}p = -1, \hat{q}^2 = +p, p^2 = -\hat{q} \right\} \tag{7.77}$$

$$\left\{ p^3 = +1, \hat{q}^3 = -1 \right\}$$

> Aside: We see that the 3-dimensional algebras are associated with the cube roots of ± 1. It is the case in general that the n-dimensional algebras are associated with the nth roots of ± 1, as will become clear later.

In the case $\left\{ p^2 = \hat{q}, \hat{p}^3 = +1 \right\}$, we have:

$$\begin{aligned}
e^{p\theta} &= 1 + p\theta + \frac{p^2\theta^2}{2!} + \frac{p^3\theta^3}{3!} + \frac{p^4\theta^4}{4!} + \frac{p^5\theta^5}{5!} + \ldots \\
&= 1 + p\theta + \frac{\hat{q}\theta^2}{2!} + \frac{\theta^3}{3!} + \frac{p\theta^4}{4!} + \frac{\hat{q}\theta^5}{5!} + \ldots \\
&= \left(1 + \frac{\theta^3}{3!} + \ldots\right) + p\left(\theta + \frac{\theta^4}{4!} + \ldots\right) + \hat{q}\left(\frac{\theta^2}{2!} + \frac{\theta^5}{5!} + \ldots\right) \\
&= AH_3(\theta) + pBH_3(\theta) + \hat{q}CH_3(\theta)
\end{aligned} \tag{7.78}$$

Similarly:

$$e^{\hat{q}\theta} = AH_3(\theta) + \hat{q}BH_3(\theta) + pCH_3(\theta) \tag{7.79}$$

This is analogous to the 2-dimensional case:

$$e^{\hat{i}\theta} = \cos\theta + \hat{i}\sin\theta \tag{7.80}$$

Clearly, there are similar identities for the other 3-dimensional flat algebras. Obscurely[96], there are similar identities for all the higher dimensional flat algebras[97]. It follows that:

$$e^{\hat{p}\frac{\theta}{n}} = AH_3\left(\frac{\theta}{n}\right) + pBH_3\left(\frac{\theta}{n}\right) + \hat{q}CH_3\left(\frac{\theta}{n}\right) \tag{7.81}$$

and:

---

[96] There is absolutely nothing in this work so far to indicate it.
[97] Your author is selling these identities for ten a penny and hopes to become infinitely rich.

Section 7

$$e^{\hat{q}\frac{\theta}{n}} = AH_3\left(\frac{\theta}{n}\right) + \hat{q}BH_3\left(\frac{\theta}{n}\right) + \hat{p}CH_3\left(\frac{\theta}{n}\right) \tag{7.82}$$

And:

$$e^{(b\hat{p}+c\hat{q})} = e^{b\hat{p}}e^{c\hat{q}}$$
$$= \left(AH_3(b) + \hat{p}BH_3(b) + \hat{q}CH_3(b)\right)\left(AH_3(c) + \hat{p}CH_3(c) + \hat{q}BH_3(c)\right)$$
$$= AH_3(b)AH_3(c) + \hat{p}AH_3(b)CH_3(c) + \hat{q}AH_3(b)BH_3(c)$$
$$+ \hat{p}BH_3(b)AH_3(c) + \hat{q}BH_3(b)CH_3(c) + BH_3(b)BH_3(c)$$
$$+ \hat{q}CH_3(b)AH_3(c) + CH_3(b)CH_3(c) + \hat{p}CH_3(b)BH_3(c)$$
$$\tag{7.83}$$

$$= v_{[c,i,n'_{(j=1,k=1)}]}A(b,c) + \hat{p}.v_{[c,i,n'_{(j=1,k=1)}]}B(b,c) + \hat{q}.v_{[c,i,n'_{(j=1,k=1)}]}C(b,c) \tag{7.84}$$

It follows that:

$$e^{\left(\frac{a+b\hat{p}+c\hat{q}}{n}\right)} = e^{\frac{a}{n}}\left(v_{[c,i,n'_{(j=1,k=1)}]}A\left(\frac{b}{n},\frac{c}{n}\right) + \hat{p}.v_{[c,i,n'_{(j=1,k=1)}]}B\left(\frac{b}{n},\frac{c}{n}\right) + \hat{q}.v_{[c,i,n'_{(j=1,k=1)}]}C\left(\frac{b}{n},\frac{c}{n}\right)\right)$$
$$\tag{7.85}$$

Hm! perhaps the simple-trig functions and the v-functions really are trigonometric functions after all![98]

The 2-dimensional Shadows of 3-dimensional Algebras:

> 2nd edition note: The concept of shadow algebras has found no use within later developments of this area of mathematics. It seems that the whole concept of a shadow algebra is a misleading distraction. We wandered from the path. None-the-less, we do not delete this idea from this 2nd edition.

In non-matrix notation, the $C_3 L^1 H^2_{(j=1,k=1)}$ algebra is the algebra of $a + b\hat{p} + c\hat{q}$ with the multiplicative relations:

$$\left\{\hat{q}p = p\hat{q} = +1,\, \hat{p}^2 = \hat{q},\hat{q}^2 = \hat{p},\, p^3 = +1, \hat{q}^3 = +1\right\} \tag{7.86}$$

In the Euclidean complex numbers algebra, the cube roots of unity have the same multiplicative relations with:

---

[98] Note how your perspicacious author is able to see the doubts within the reader's mind. Perhaps it is because he too once harbored those doubts.

$$\left\{p = -\frac{1}{2} + \frac{\sqrt{3}}{2}\hat{i}, \hat{q} = -\frac{1}{2} - \frac{\sqrt{3}}{2}\hat{i}\right\} \tag{7.87}$$

If we substitute these complex roots into the non-matrix representation of the $C_3L^1H^2_{(j=1,k=1)}$ algebra, we get:

$$\begin{aligned} a + bp + c\hat{q} &= a + b\left(-\frac{1}{2} + \frac{\sqrt{3}}{2}\hat{i}\right) + c\left(-\frac{1}{2} - \frac{\sqrt{3}}{2}\hat{i}\right) \\ &= \left(a - \frac{b+c}{2}\right) + \hat{i}\left(\frac{\sqrt{3}}{2}(b-c)\right) \end{aligned} \tag{7.88}$$

When we reduce the $C_3L^1H^2_{(j=1,k=1)}$ algebra to complex numbers, we are squeezing a 3-dimensional algebra into a 2-dimensional representation of that algebra. We say that the 2-dimensional representation is a 2-dimensional shadow of the 3-dimensional algebra.

Nomenclature:
Shadow Algebra
A shadow algebra is a lesser dimensional representation of an algebra.

We still have three independent variables $\{a, b, c\}$ in the complex shadow of the $C_3L^1H^2_{(j=1,k=1)}$ algebra. The complex shadow is an algebraic field for all values of the three variables – it is the complex numbers. All natural algebraic fields are associated with a natural space. Thus, we are shadowing the 3-dimensional space of the $C_3L^1H^2_{(j=1,k=1)}$ algebra in the 2-dimensional space of the Euclidean complex numbers - 2-dimensional Euclidean space.

Physical 2-dimensional shadows of 3-dimensional objects retain some of the detail of the 3-dimensional object but also lose some detail. So it is with shadow algebras and shadow spaces. The distance function of the $C_3L^1H^2_{(j=1,k=1)}$ algebra is based on the determinant of the matrix form of that algebra, which is:

$$a^3 + b^3 + c^3 - 3abc = (a+b+c)(a^2 + b^2 + c^2 - ab - ac - bc) \tag{7.89}$$

The shadow algebra, being a natural algebra, has, within it, a distance function based on the determinant, which is:

Section 7

$$\left(a-\frac{(b+c)}{2}\right)^2 + \left(\frac{\sqrt{3}}{2}(b-c)\right)^2 = \left(a^2+b^2+c^2-ab-ac-bc\right) \quad (7.90)$$

We have lost the factor $(a+b+c)$. This factor is of the form of the 1-dimensional metric – a linear factor – we seem to have squashed out one dimension.

The reader should realise that we are not cutting a 2-dimensional sub-space out of the 3-dimensional space[99]; we are representing the whole of the 3-dimensional space as a 2-dimensional shadow of itself.

The polar form of the shadow representation is:

$$\exp\left(\begin{bmatrix} a-\frac{(b+c)}{2} & \frac{\sqrt{3}}{2}(b-c) \\ -\frac{\sqrt{3}}{2}(b-c) & a-\frac{(b+c)}{2} \end{bmatrix}\right)$$

$$= \exp\left(\begin{bmatrix} a & 0 \\ 0 & a \end{bmatrix} + \begin{bmatrix} -\frac{(b+c)}{2} & 0 \\ 0 & -\frac{(b+c)}{2} \end{bmatrix} + \begin{bmatrix} 0 & \frac{\sqrt{3}}{2}(b-c) \\ -\frac{\sqrt{3}}{2}(b-c) & 0 \end{bmatrix}\right)$$

(7.91)

$$= \begin{bmatrix} e^a & 0 \\ 0 & e^a \end{bmatrix} \begin{bmatrix} e^{-\frac{(b+c)}{2}} & 0 \\ 0 & e^{-\frac{(b+c)}{2}} \end{bmatrix} \begin{bmatrix} \cos\left(\frac{\sqrt{3}}{2}(b-c)\right) & \sin\left(\frac{\sqrt{3}}{2}(b-c)\right) \\ -\sin\left(\frac{\sqrt{3}}{2}(b-c)\right) & \cos\left(\frac{\sqrt{3}}{2}(b-c)\right) \end{bmatrix}$$

(7.92)

We see that the length within the 3-dimensional algebra, e^a, has now become $e^{a-\frac{b+c}{2}}$. The 3-dimensional algebras have two rotation matrices in the simple polar form. When we combine these together to form the compound polar form, we get a single rotation matrix containing the v-functions. The reader is urged to compare the above shadow with the standard form of the v-functions for the $C_3L^1H^2_{(j=1,k=1)}$ algebra. Analogous results occur for the three other 3-dimensional algebras.

---

[99] 3-dimensional natural spaces do not have 2-dimensional sub-spaces, but that's jumping the gun.

Logarithms - $C_3L^1H^2_{(j=1,k=1)}$:

The simple polar form of the $C_3L^1H^2_{(j=1,k=1)}$ algebra is:

$$\begin{bmatrix} a & b & c \\ c & a & b \\ b & c & a \end{bmatrix} = PROD \begin{cases} \begin{bmatrix} h & 0 & 0 \\ 0 & h & 0 \\ 0 & 0 & h \end{bmatrix} \begin{bmatrix} AH_3(\beta) & BH_3(\beta) & CH_3(\beta) \\ CH_3(\beta) & AH_3(\beta) & BH_3(\beta) \\ BH_3(\beta) & CH_3(\beta) & AH_3(\beta) \end{bmatrix} \\ \begin{bmatrix} AH_3(\chi) & CH_3(\chi) & BH_3(\chi) \\ BH_3(\chi) & AH_3(\chi) & CH_3(\chi) \\ CH_3(\chi) & BH_3(\chi) & AH_3(\chi) \end{bmatrix} \end{cases} \quad (7.93)$$

Thus:

$$\log m_{[e]} \left(\begin{bmatrix} a & b & c \\ c & a & b \\ b & c & a \end{bmatrix} \right)$$

$$= \log m_{[e]} \left(PROD \begin{cases} \begin{bmatrix} h & 0 & 0 \\ 0 & h & 0 \\ 0 & 0 & h \end{bmatrix} \begin{bmatrix} AH_3(\beta) & BH_3(\beta) & CH_3(\beta) \\ CH_3(\beta) & AH_3(\beta) & BH_3(\beta) \\ BH_3(\beta) & CH_3(\beta) & AH_3(\beta) \end{bmatrix} \\ \begin{bmatrix} AH_3(\chi) & CH_3(\chi) & BH_3(\chi) \\ BH_3(\chi) & AH_3(\chi) & CH_3(\chi) \\ CH_3(\chi) & BH_3(\chi) & AH_3(\chi) \end{bmatrix} \end{cases} \right) \quad (7.94)$$

$$= \log m_{[e]} \left(\begin{bmatrix} h & 0 & 0 \\ 0 & h & 0 \\ 0 & 0 & h \end{bmatrix} \right) + \log m_{[e]} \left(\begin{bmatrix} AH_3(\beta) & BH_3(\beta) & CH_3(\beta) \\ CH_3(\beta) & AH_3(\beta) & BH_3(\beta) \\ BH_3(\beta) & CH_3(\beta) & AH_3(\beta) \end{bmatrix} \right)$$

$$+ \log m_{[e]} \left(\begin{bmatrix} AH_3(\chi) & CH_3(\chi) & BH_3(\chi) \\ BH_3(\chi) & AH_3(\chi) & CH_3(\chi) \\ CH_3(\chi) & BH_3(\chi) & AH_3(\chi) \end{bmatrix} \right) \quad (7.95)$$

Section 7

$$= \log m_{[e]} \left(\begin{bmatrix} h & 0 & 0 \\ 0 & h & 0 \\ 0 & 0 & h \end{bmatrix} \right) + \log m_{[e]} \left(\exp \begin{bmatrix} 0 & \beta & 0 \\ 0 & 0 & \beta \\ \beta & 0 & 0 \end{bmatrix} \right)$$

$$+ \log m_{[e]} \left(\exp \begin{bmatrix} 0 & 0 & \chi \\ \chi & 0 & 0 \\ 0 & \chi & 0 \end{bmatrix} \right)$$

$$= \log m_{[e]} \left(\begin{bmatrix} h & 0 & 0 \\ 0 & h & 0 \\ 0 & 0 & h \end{bmatrix} \right) + \begin{bmatrix} 0 & \beta & 0 \\ 0 & 0 & \beta \\ \beta & 0 & 0 \end{bmatrix} + \begin{bmatrix} 0 & 0 & \chi \\ \chi & 0 & 0 \\ 0 & \chi & 0 \end{bmatrix}$$

$$= \begin{bmatrix} \log_e h & \beta & \chi \\ \chi & \log_e h & \beta \\ \beta & \chi & \log_e h \end{bmatrix}$$

(7.96)

In non-matrix notation, this is:

$$\log_e (a + bp + c\hat{q}) = \log_e h + \beta p + \chi \hat{q}$$ (7.97)

3-dimensional Vectors:

We take two vectors from the $C_3 L^1 H^2_{(j=1,k=1)}$ algebra:

$$\left\{ \begin{bmatrix} a & b & c \\ c & a & b \\ b & c & a \end{bmatrix}, \begin{bmatrix} d & e & f \\ f & d & e \\ e & f & d \end{bmatrix} \right\}$$ (7.98)

We normalise them:

$$\left\{ \frac{1}{\sqrt[3]{a^3 + b^3 + c^3 - 3abc}} \begin{bmatrix} a & b & c \\ c & a & b \\ b & c & a \end{bmatrix}, \right.$$

$$\left. \frac{1}{\sqrt[3]{d^3 + e^3 + f^3 - 3def}} \begin{bmatrix} d & e & f \\ f & d & e \\ e & f & d \end{bmatrix} \right\}$$ (7.99)

We take the conjugate of one (either) of the normalised matrices:

$$\begin{bmatrix} \dfrac{a^2-bc}{\left(\sqrt[3]{a^3+b^3+c^3-3abc}\right)^2} & \dfrac{ab-c^2}{\left(\sqrt[3]{a^3+b^3+c^3-3abc}\right)^2} & \dfrac{ac-b^2}{\left(\sqrt[3]{a^3+b^3+c^3-3abc}\right)^2} \\ \dfrac{ac-b^2}{\left(\sqrt[3]{a^3+b^3+c^3-3abc}\right)^2} & \dfrac{a^2-bc}{\left(\sqrt[3]{a^3+b^3+c^3-3abc}\right)^2} & \dfrac{ab-c^2}{\left(\sqrt[3]{a^3+b^3+c^3-3abc}\right)^2} \\ \dfrac{ab-c^2}{\left(\sqrt[3]{a^3+b^3+c^3-3abc}\right)^2} & \dfrac{ac-b^2}{\left(\sqrt[3]{a^3+b^3+c^3-3abc}\right)^2} & \dfrac{a^2-bc}{\left(\sqrt[3]{a^3+b^3+c^3-3abc}\right)^2} \end{bmatrix}$$

(7.100)

We note that the normalisation factor is squared. We multiply this conjugate by the other normalised matrix to get:

$$\begin{bmatrix} \dfrac{a^2d+b^2e+c^2f-abf-ace-bcd}{\left(\sqrt[3]{a^3+b^3+c^3-3abc}\right)^2 \sqrt[3]{d^3+e^3+f^3-3def}} & \sim & \sim \\ \dfrac{a^2f+b^2d+c^2e-abe-acd-bcf}{\left(\sqrt[3]{a^3+b^3+c^3-3abc}\right)^2 \sqrt[3]{d^3+e^3+f^3-3def}} & \sim & \sim \\ \dfrac{a^2e+b^2f+c^2d-abd-acf-bce}{\left(\sqrt[3]{a^3+b^3+c^3-3abc}\right)^2 \sqrt[3]{d^3+e^3+f^3-3def}} & \sim & \sim \end{bmatrix}$$

(7.101)

Leading to:

$$v_{[\,]}A(\chi,\theta) = \dfrac{a^2d+b^2e+c^2f-abf-ace-bcd}{\left(\sqrt[3]{a^3+b^3+c^3-3abc}\right)^2 \sqrt[3]{d^3+e^3+f^3-3def}}$$

$$v_{[\,]}B(\chi,\theta) = \dfrac{a^2e+b^2f+c^2d-abd-acf-bce}{\left(\sqrt[3]{a^3+b^3+c^3-3abc}\right)^2 \sqrt[3]{d^3+e^3+f^3-3def}}$$

(7.102)

$$v_{[\,]}C(\chi,\theta) = \dfrac{a^2f+b^2d+c^2e-abe-acd-bcf}{\left(\sqrt[3]{a^3+b^3+c^3-3abc}\right)^2 \sqrt[3]{d^3+e^3+f^3-3def}}$$

In the case that the two vectors are the same vector:

Section 7

$$v_{[\,]}A(\chi,\theta) = 1$$
$$v_{[\,]}B(\chi,\theta) = 0 \qquad (7.103)$$
$$v_{[\,]}C(\chi,\theta) = 0$$

In this case, $\theta = \chi = 0$, of course.

Section 8

Introduction to Section 8:

In this section, we examine the 3-dimensional trigonometric functions more deeply. Much of this section is a catalogue of series and differential cycles.

After the trigonometric functions have been examined more closely, we come to a result regarding the algebraic closure of the natural algebras. This result is a deep insight into the nature of algebraic closure. The result is that the Euclidean complex numbers are the only natural algebra that is algebraically closed. Actually, natural algebras that contain the Euclidean complex numbers as a sub-algebra are also algebraically closed, but that is only because they include the Euclidean complex numbers.

The 3-dimensional Simple-Trig Functions:

We have previously met the 3-dimensional simple-trig functions in the simple polar forms of the 3-dimensional flat algebras. We think of the 3-dimensional simple-trig functions as being projection functions that project from the unit sphere on to a 2-dimensional plane formed by two axes.

We defined the 3-dimensional simple-trig functions as 3-ways splittings of the exponential series (with some minus signs chucked in for the E-type). Those definitions are:

$$AH_3(x) = hypergeom\left([\ ], \left[\frac{1}{3}, \frac{2}{3}\right], \left(\frac{x}{3}\right)^3\right) = \frac{x^0}{0!} + \frac{x^3}{3!} + \frac{x^6}{6!} + \frac{x^9}{9!} + \ldots$$

$$BH_3(x) = \frac{x}{1!}\left(hypergeom\left([\ ], \left[\frac{2}{3}, \frac{4}{3}\right], \left(\frac{x}{3}\right)^3\right)\right) = \frac{x^1}{1!} + \frac{x^4}{4!} + \frac{x^7}{7!} + \frac{x^{10}}{10!} + \ldots$$

$$CH_3(x) = \frac{x^2}{2!}\left(hypergeom\left([\ ], \left[\frac{4}{3}, \frac{5}{3}\right], \left(\frac{x}{3}\right)^3\right)\right) = \frac{x^2}{2!} + \frac{x^5}{5!} + \frac{x^8}{8!} + \frac{x^{11}}{11!} + \ldots$$

(8.1)

$$AE_3(x) = hypergeom\left([\ ], \left[\frac{1}{3}, \frac{2}{3}\right], -\left(\frac{x}{3}\right)^3\right) = \frac{x^0}{0!} - \frac{x^3}{3!} + \frac{x^6}{6!} - \frac{x^9}{9!} + \ldots$$

$$BE_3(x) = \frac{x}{1!}\left(hypergeom\left([\ ], \left[\frac{2}{3}, \frac{4}{3}\right], -\left(\frac{x}{3}\right)^3\right)\right) = \frac{x^1}{1!} - \frac{x^4}{4!} + \frac{x^7}{7!} - \frac{x^{10}}{10!} + \ldots$$

(8.2)

Section 8

$$CE_3(x) = \frac{x^2}{2!}\left(\text{hypergeom}\left([\ ],\left[\frac{4}{3},\frac{5}{3}\right],-\left(\frac{x}{3}\right)^3\right)\right) = \frac{x^2}{2!} - \frac{x^5}{5!} + \frac{x^8}{8!} - \frac{x^{11}}{11!} + \ldots$$
(8.3)

The first letter of the appellations chosen is just a letter. The second letter is either an H to denote H-type or an E to denote E-type. The sub-script number is the splitting of the exponential; it is also the dimension of the space of the algebra associated with a particular set of simple-trig functions. Using the above nomenclature, the familiar 2-dimensional trigonometric functions would be:

$$\{\cos \equiv AE_2,\ \sin \equiv BE_2,\ \cosh \equiv AH_2,\ \sinh \equiv BH_2\}$$
(8.4)

and the 1-dimensional trigonometric function, 1, might be written as AH_1.

We immediately have:
$$e^x = AH_3(x) + BH_3(x) + CH_3(x)$$
$$e^{-x} = AE_3(x) - BE_3(x) + CE_3(x)$$
(8.5)

Hence:
$$(AH_3(x) + BH_3(x) + CH_3(x))(AE_3(x) - BE_3(x) + CE_3(x)) = 1$$
(8.6)

Other identities are:
$$AE_3(x)BE_3(x) - AE_3(x)CE_3(x) + BE_3(x)CE_3(x) = \frac{1}{3}\left(-e^{-2x} + e^x\right)$$

$$AH_3(x)BH_3(x) + AH_3(x)CH_3(x) + BH_3(x)CH_3(x) = \frac{1}{3}\left(e^{2x} - e^{-x}\right)$$
(8.7)

We also have:
$$AH_3(-x) = AE_3(x)$$
$$BH_3(-x) = -BE_3(x)$$
(8.8)
$$CH_3(-x) = CE_3(x)$$
$$AE_3(-x) = AH_3(x)$$
$$BE_3(-x) = -BH_3(x)$$
(8.9)
$$CE_3(-x) = CH_3(x)$$

Differentiation:

Clearly[100], we have cyclic differential relations similar to *sinh* & *cosh* and *sin* & *cos* but with a 6 & 3-cycle rather than a 4 & 2-cycle:

$$\begin{cases} \dfrac{d}{dx}CE_3x = BE_3x, & \dfrac{d^2}{dx^2}CE_3x = AE_3x, & \dfrac{d^3}{dx^3}CE_3x = -CE_3x, \\ \dfrac{d^4}{dx^4}CE_3x = -BE_3x, & \dfrac{d^5}{dx^5}CE_3x = -AE_3x, & \dfrac{d^6}{dx^6}CE_3x = CE_3x \end{cases} \quad (8.10)$$

$$\left\{ \dfrac{d}{dx}CH_3x = BH_3x, \quad \dfrac{d^2}{dx^2}CH_3x = AH_3x, \quad \dfrac{d^3}{dx^3}CH_3x = CH_3x \right\}$$

In Standard Function Form:

These simple-trig functions have representation as standard functions:

$$AH_3(x) = \frac{1}{3}\left(e^x + 2e^{-\left(\frac{x}{2}\right)} \cos\left(\frac{\sqrt{3}}{2}x\right) \right)$$

$$BH_3(x) = \frac{1}{3}\left(e^x + e^{-\left(\frac{x}{2}\right)}\left(\sqrt{3}\sin\left(\frac{\sqrt{3}}{2}x\right) - \cos\left(\frac{\sqrt{3}}{2}x\right) \right) \right) \quad (8.11)$$

$$CH_3(x) = \frac{1}{3}\left(e^x - e^{-\left(\frac{x}{2}\right)}\left(\sqrt{3}\sin\left(\frac{\sqrt{3}}{2}x\right) + \cos\left(\frac{\sqrt{3}}{2}x\right) \right) \right)$$

$$AE_3(x) = \frac{1}{3}\left(e^{-x} + 2e^{\left(\frac{x}{2}\right)} \cos\left(\frac{\sqrt{3}}{2}x\right) \right)$$

$$BE_3(x) = \frac{1}{3}\left(-e^{-x} + e^{\left(\frac{x}{2}\right)}\left(\sqrt{3}\sin\left(\frac{\sqrt{3}}{2}x\right) + \cos\left(\frac{\sqrt{3}}{2}x\right) \right) \right) \quad (8.12)$$

$$CE_3(x) = \frac{1}{3}\left(e^{-x} + e^{\left(\frac{x}{2}\right)}\left(\sqrt{3}\sin\left(\frac{\sqrt{3}}{2}x\right) - \cos\left(\frac{\sqrt{3}}{2}x\right) \right) \right)$$

We have:

$$2\cos\left(\frac{\sqrt{3}}{2}x + \frac{2\pi}{3}\right) = -\left(\sqrt{3}\sin\left(\frac{\sqrt{3}}{2}x\right) + \cos\left(\frac{\sqrt{3}}{2}x\right) \right)$$

$$2\cos\left(\frac{\sqrt{3}}{2}x + \frac{\pi}{3}\right) = -\left(\sqrt{3}\sin\left(\frac{\sqrt{3}}{2}x\right) - \cos\left(\frac{\sqrt{3}}{2}x\right) \right)$$

(8.13)

Giving:

---

[100] It is clear when you think of the series representation.

Section 8

$$AH_3(x) = \frac{1}{3}\left(e^x + 2e^{-\left(\frac{x}{2}\right)}\cos\left(\frac{\sqrt{3}}{2}x\right)\right)$$

$$BH_3(x) = \frac{1}{3}\left(e^x - 2e^{-\left(\frac{x}{2}\right)}\cos\left(\frac{\sqrt{3}}{2}x + \frac{\pi}{3}\right)\right) \tag{8.14}$$

$$CH_3(x) = \frac{1}{3}\left(e^x + 2e^{-\left(\frac{x}{2}\right)}\cos\left(\frac{\sqrt{3}}{2}x + \frac{2\pi}{3}\right)\right)$$

And:

$$AE_3(x) = \frac{1}{3}\left(e^{-x} + 2e^{\left(\frac{x}{2}\right)}\cos\left(\frac{\sqrt{3}}{2}x\right)\right)$$

$$BE_3(x) = -\frac{1}{3}\left(e^{-x} + 2e^{\left(\frac{x}{2}\right)}\cos\left(\frac{\sqrt{3}}{2}x + \frac{2\pi}{3}\right)\right) \tag{8.15}$$

$$CE_3(x) = \frac{1}{3}\left(e^{-x} - 2e^{\left(\frac{x}{2}\right)}\cos\left(\frac{\sqrt{3}}{2}x + \frac{\pi}{3}\right)\right)$$

Graphs:

When $x = 0$:

$$AH_3(0) = 1 \quad BH_3(0) = 0 \quad CH_3(0) = 0 \tag{8.16}$$

AH_3 has a point of inflection at $x = 0$. CH_3 has a local minimum at $x = 0$. BH_3 passes through 0 with slope 1. These functions are wave-like for small values of x. They are wave-like with enormous amplitude for large negative x, but, for large positive x, the wave-like nature is swamped and they appear non-wavelike unless examined minutely.

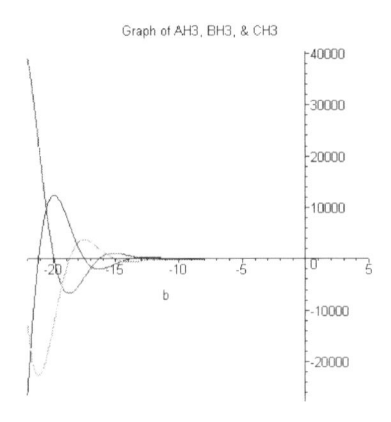

Graph of AH3, BH3, & CH3

Graph of AH3, BH3, & CH3

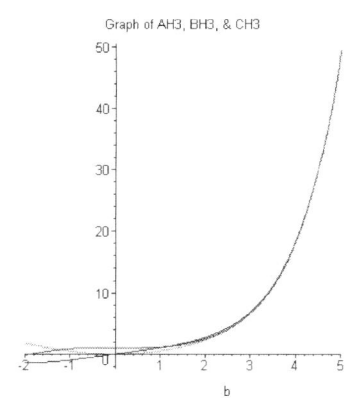

The wavy parts of the 3-dimensional simple-trig functions are out of phase with each other by $\frac{\pi}{3}$. However, the presence of the exponential parts of these functions means that they cannot be translated along the *x*-axis to sit on top of each other in the way the 2-dimensional Euclidean trigonometric functions can.

When $x = 0$:

$$AE_3(0) = 1$$
$$BE_3(0) = 0 \qquad (8.17)$$
$$CE_3(0) = 0$$

AE_3 has a point of inflection at $x = 0$. CE_3 has a local minimum at $x = 0$. BE_3 passes through 0 with slope 1. These functions are wave-like with enormous amplitude for large positive values of *x*. For large negative *x*, the wave-like nature is swamped and they appear non-wavelike unless examined minutely. BE_3 grows exponentially negative while AE_3 and CE_3 grow exponentially positive.

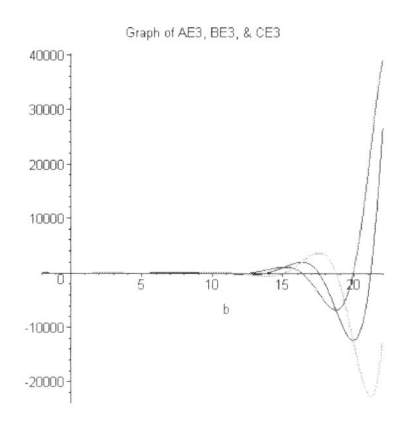

Graph of AE3, BE3, & CE3

Section 8

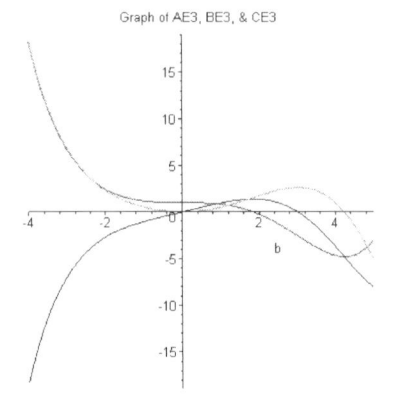

Graph of AE3, BE3, & CE3

It is left as an exercise for the reader to do Fourier analysis with these simple-trig functions.

The $C_3L^1H^2_{(j=1,k=1)}$ V-Functions:

We have met the 3-dimensional v-functions previously. Each of the four 3-dimensional flat algebras has a set of v-functions. Although it is possible to construct a general 3-dimensional v-function from the general 3-dimensional polar form, the expression is very complicated and we choose not to include it in this work.

The 3-dimensional v-functions are projections from the unit sphere on to the axes of the space.

The $C_3L^1H^2_{(j=1,k=1)}$ v-functions are:

$$v_{[C_3L^1H^2_{(j=1,k=1)}]}A(b,c) = \frac{1}{3}\left(e^{(b+c)} + 2e^{-\left(\frac{b+c}{2}\right)}\cos\left(\frac{\sqrt{3}}{2}(b-c)\right)\right)$$

$$v_{[C_3L^1H^2_{(j=1,k=1)}]}B(b,c) = \frac{1}{3}\left(e^{(b+c)} - 2e^{-\left(\frac{b+c}{2}\right)}\cos\left(\frac{\sqrt{3}}{2}(b-c) + \frac{\pi}{3}\right)\right) \quad (8.18)$$

$$v_{[C_3L^1H^2_{(j=1,k=1)}]}C(b,c) = \frac{1}{3}\left(e^{(b+c)} + 2e^{-\left(\frac{b+c}{2}\right)}\cos\left(\frac{\sqrt{3}}{2}(b-c) + \frac{2\pi}{3}\right)\right)$$

When we set one of the variables to zero, we reduce the v-functions to the simple-trig functions. When $b = 0$:

$$v_{\left[C_3l^1H^2_{(J-1,k-1)}\right]}A(0,c) = AH_3(c)$$
$$v_{\left[C_3l^1H^2_{(J-1,k-1)}\right]}B(0,c) = CH_3(c) \qquad (8.19)$$
$$v_{\left[C_3l^1H^2_{(J-1,k-1)}\right]}C(0,c) = BH_3(c)$$

When $c = 0$:
$$v_{\left[C_3l^1H^2_{(J-1,k-1)}\right]}A(b,0) = AH_3(b)$$
$$v_{\left[C_3l^1H^2_{(J-1,k-1)}\right]}B(b,0) = BH_3(b) \qquad (8.20)$$
$$v_{\left[C_3l^1H^2_{(J-1,k-1)}\right]}C(b,0) = CH_3(b)$$

And:
$$v_{\left[C_3l^1H^2_{(J-1,k-1)}\right]}A(0,0) = 1$$
$$v_{\left[C_3l^1H^2_{(J-1,k-1)}\right]}B(0,0) = 0 \qquad (8.21)$$
$$v_{\left[C_3l^1H^2_{(J-1,k-1)}\right]}C(0,0) = 0$$

We have:
$$v_{\left[C_3l^1H^2_{(J-1,k-1)}\right]}A(b,c) + v_{\left[C_3l^1H^2_{(J-1,k-1)}\right]}B(b,c) + v_{\left[C_3l^1H^2_{(J-1,k-1)}\right]}C(b,c) = e^{(b+c)} \qquad (8.22)$$

but the v-functions are not simple splittings of the exponential series of e^{b+c}. It is the simple-trig functions that are the simple splittings of the exponential series.

Differentiation:

We have:
$$\frac{\partial\left(v_{\left[C_3l^1H^2_{(J-1,k-1)}\right]}A(b,c)\right)}{\partial b} = v_{\left[C_3l^1H^2_{(J-1,k-1)}\right]}C(b,c)$$
$$\frac{\partial\left(v_{\left[C_3l^1H^2_{(J-1,k-1)}\right]}A(b,c)\right)}{\partial c} = v_{\left[C_3l^1H^2_{(J-1,k-1)}\right]}B(b,c) \qquad (8.23)$$

$$\frac{\partial\left(v_{\left[C_3l^1H^2_{(J-1,k-1)}\right]}B(b,c)\right)}{\partial b} = v_{\left[C_3l^1H^2_{(J-1,k-1)}\right]}A(b,c) \qquad (8.24)$$

Section 8

$$\frac{\partial\left(v_{[C_3L'H^2_{(j-1,k-1)}]}B(b,c)\right)}{\partial b} = v_{[C_3L'H^2_{(j-1,k-1)}]}A(b,c)$$

$$\frac{\partial\left(v_{[C_3L'H^2_{(j-1,k-1)}]}B(b,c)\right)}{\partial c} = v_{[C_3L'H^2_{(j-1,k-1)}]}C(b,c)$$
(8.25)

$$\frac{\partial\left(v_{[C_3L'H^2_{(j-1,k-1)}]}C(b,c)\right)}{\partial b} = v_{[C_3L'H^2_{(j-1,k-1)}]}B(b,c)$$

$$\frac{\partial\left(v_{[C_3L'H^2_{(j-1,k-1)}]}C(b,c)\right)}{\partial c} = v_{[C_3L'H^2_{(j-1,k-1)}]}A(b,c)$$
(8.26)

Thus we have a differentiation cycle.

Anything with a repeating differentiation cycle just has[101] to be something to do with the exponential series. We have:

Expanding about $c = 0$:

$$v_{[C_3L'H^2_{(j-1,k-1)}]}A(b,c) = \frac{[AH_3(b)]c^0}{0!} + \frac{[BH_3(b)]c^1}{1!} + \frac{[CH_3(b)]c^2}{2!} + \frac{[AH_3(b)]c^3}{3!} + \ldots$$

$$= \sum_{n=0}^{\infty}\left(\frac{[AH_3(b)]c^{3n}}{(3n)!} + \frac{[BH_3(b)]c^{3n+1}}{(3n+1)!} + \frac{[CH_3(b)]c^{3n+2}}{(3n+2)!}\right)$$
(8.27)

Inspection of this reveals that differentiation with respect to either variable (3 times) leads to the same series. How remarkably intimate these functions are!

Similarly:

Expanding about $b = 0$:

$$v_{[C_3L'H^2_{(j-1,k-1)}]}A(b,c) = \frac{[AH_3(c)]b^0}{0!} + \frac{[BH_3(c)]b^1}{1!} + \frac{[CH_3(c)]b^2}{2!} + \frac{[AH_3(c)]b^3}{3!} + \ldots$$
(8.28)

Expanding about $c = 0$:

$$v_{[C_3L'H^2_{(j-1,k-1)}]}B(b,c) = \frac{[BH_3(b)]c^0}{0!} + \frac{[CH_3(b)]c^1}{1!} + \frac{[AH_3(b)]c^2}{2!} + \frac{[BH_3(b)]c^3}{3!} + \ldots$$
(8.29)

Expanding about $b = 0$:

---

[101] This is an emotional outburst not a mathematical fact.

$$v_{[C_3 l^i H^2_{(j-1,k-1)}]} B(b,c) = \frac{[CH_3(c)]b^0}{0!} + \frac{[AH_3(c)]b^1}{1!} + \frac{[BH_3(c)]b^2}{2!} + \frac{[CH_3(c)]b^3}{3!} + \dots$$
(8.30)

Expanding about $c = 0$:

$$v_{[C_3 l^i H^2_{(j-1,k-1)}]} C(b,c) = \frac{[CH_3(b)]c^0}{0!} + \frac{[AH_3(b)]c^1}{1!} + \frac{[BH_3(b)]c^2}{2!} + \frac{[CH_3(b)]c^3}{3!} + \dots$$
(8.31)

Expanding about $b = 0$:

$$v_{[C_3 l^i H^2_{(j-1,k-1)}]} C(b,c) = \frac{[BH_3(c)]b^0}{0!} + \frac{[CH_3(c)]b^1}{1!} + \frac{[AH_3(c)]b^2}{2!} + \frac{[BH_3(c)]b^3}{3!} + \dots$$
(8.32)

3D-Graphs:

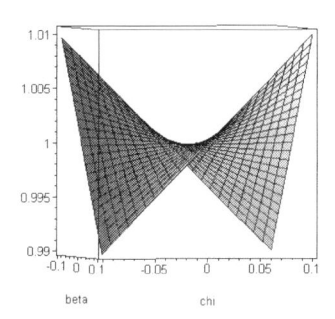

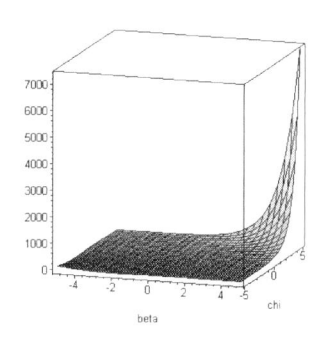

All three function plotted in the same axes are a tsunami graph for small and negative values of b and c, but, for large positive values of b and c, the functions join together to race off exponentially toward infinity.

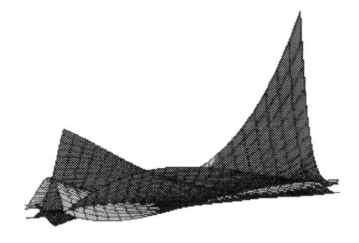

Similar results apply to the other 3-dimensional algebras.

Section 9

Section 9

Introduction to Section 9:

In this section, we meet the 4-dimensional natural algebras and the 4-dimensional simple-trig functions. Much within this section is analogous to the 3-dimensional natural algebras, but there are some differences.

There is only one basic 3-dimensional algebraic matrix form. There are three plus one basic algebraic matrix forms in four dimensions. The 3-dimensional natural algebras contain only the real numbers as a sub-algebra. The 4-dimensional natural algebras contain, as well as the real numbers, 2-dimensional sub-algebras.

The Polar Form of the $C_{4.1} L^1 H^3_{(j=1,k=1,l=1)}$ Algebra:

The general form of the $C_{4.1}$ algebra is:

$$C_{4.1} = \begin{bmatrix} a & b & c & d \\ jd & a & kb & lc \\ \dfrac{jl}{k}c & \dfrac{j}{k}d & a & lb \\ jb & \dfrac{j}{k}c & \dfrac{j}{l}d & a \end{bmatrix} \qquad (9.1)$$

We take $\{j=1, k=1, l=1\}$ and exponentiate this matrix.

2nd edition note: Your author originally hoped to find our 4-dimensional space-time in the 4-dimensional algebras shown here. The algebras that derive from the C_4 group are all commutative and seem to play no part in the physical universe. The commutative algebras from the group $C_2 \times C_2$ also seem to play no part in the physical universe. These are the only 4-doimensional algebras considered in this book. The eight non-commutative algebras within the group $C_2 \times C_2$ contain the space-time of our universe together with the whole of General Relativity and classical electromagnetism. They might also contain the weak force. Missed 'em!

$$\exp\left(\begin{bmatrix} a & b & c & d \\ d & a & b & c \\ c & d & a & b \\ b & c & d & a \end{bmatrix}\right) = PROD \left(\begin{bmatrix} e^a & 0 & 0 & 0 \\ 0 & e^a & 0 & 0 \\ 0 & 0 & e^a & 0 \\ 0 & 0 & 0 & e^a \end{bmatrix} \begin{bmatrix} AH_4(b) & BH_4(b) & CH_4(b) & DH_4(b) \\ DH_4(b) & AH_4(b) & BH_4(b) & CH_4(b) \\ CH_4(b) & DH_4(b) & AH_4(b) & BH_4(b) \\ BH_4(b) & CH_4(b) & DH_4(b) & AH_4(b) \end{bmatrix} \begin{bmatrix} \cosh(c) & 0 & \sinh(c) & 0 \\ 0 & \cosh(c) & 0 & \sinh(c) \\ \sinh(c) & 0 & \cosh(c) & 0 \\ 0 & \sinh(c) & 0 & \cosh(c) \end{bmatrix} \begin{bmatrix} AH_4(d) & DH_4(d) & CH_4(d) & BH_4(d) \\ BH_4(d) & AH_4(d) & DH_4(d) & CH_4(d) \\ CH_4(d) & BH_4(d) & AH_4(d) & DH_4(d) \\ DH_4(d) & CH_4(d) & BH_4(d) & AH_4(d) \end{bmatrix}\right)$$

(9.2)

Where:

$$AH_4(x) = 1 + \frac{x^4}{4!} + \frac{x^8}{8!} + \dots$$

$$BH_4(x) = x + \frac{x^5}{5!} + \frac{x^9}{9!} + \dots$$

$$CH_4(x) = \frac{x^2}{2!} + \frac{x^6}{6!} + \frac{x^{10}}{10!} + \dots$$

$$DH_4(x) = \frac{x^3}{3!} + \frac{x^7}{7!} + \frac{x^{11}}{11!} + \dots$$

(9.3)

are the 4-way splittings of the exponential series that are the 4-dimensional simple-trig functions.

The matrices containing the simple-trig functions were formed by exponentiating a matrix with trace zero and thus the determinants of these matrices are unity. Thereby do we have a trigonometric identity:

Section 9

$$AH_4^{\,4}(x) - BH_4^{\,4}(x) + CH_4^{\,4}(x) - DH_4^{\,4}(x) - 2AH_4^{\,2}(x)CH_4^{\,2}(x)$$
$$+2BH_4^{\,2}(x)DH_4^{\,2}(x) - 4AH_4^{\,2}(x)BH_4(x)DH_4(x)$$
$$+4AH_4(x)BH_4^{\,2}(x)CH_4(x) - 4BH_4(x)CH_4^{\,2}(x)DH_4(x)$$
$$+4AH_4(x)CH_4(x)DH_4^{\,2}(x) = 1$$
(9.4)

This expression will factorise:

$$\begin{pmatrix} (AH_4(x) + BH_4(x) + CH_4(x) + DH_4(x)) \\ (AH_4(x) - BH_4(x) + CH_4(x) - DH_4(x)) \\ ((BH_4(x) - DH_4(x))^2 + (AH_4(x) - CH_4(x))^2) \end{pmatrix} = 1 \quad (9.5)$$

Consideration of the series expansions leads to:

$$(e^x)(e^{-x})(\sin^2(x) + \cos^2(x)) = 1 \quad (9.6)$$

Which is, of course, well-known. We also have:

$$\begin{bmatrix} AH_4(b+x) & BH_4(b+x) & CH_4(b+x) & DH_4(b+x) \\ DH_4(b+x) & AH_4(b+x) & BH_4(b+x) & CH_4(b+x) \\ CH_4(b+x) & DH_4(b+x) & AH_4(b+x) & BH_4(b+x) \\ BH_4(b+x) & CH_4(b+x) & DH_4(b+x) & AH_4(b+x) \end{bmatrix}$$

$$= \text{PROD} \begin{pmatrix} \begin{bmatrix} AH_4(b) & BH_4(b) & CH_4(b) & DH_4(b) \\ DH_4(b) & AH_4(b) & BH_4(b) & CH_4(b) \\ CH_4(b) & DH_4(b) & AH_4(b) & BH_4(b) \\ BH_4(b) & CH_4(b) & DH_4(b) & AH_4(b) \end{bmatrix} \\ \begin{bmatrix} AH_4(x) & BH_4(x) & CH_4(x) & DH_4(x) \\ DH_4(x) & AH_4(x) & BH_4(x) & CH_4(x) \\ CH_4(x) & DH_4(x) & AH_4(x) & BH_4(x) \\ BH_4(x) & CH_4(x) & DH_4(x) & AH_4(x) \end{bmatrix} \end{pmatrix} \quad (9.7)$$

Leading to the trigonometric identities:

$$AH_4(b+x) = AH_4(b)AH_4(x) + BH_4(b)DH_4(x)$$
$$+DH_4(b)BH_4(x) + CH_4(b)CH_4(x) \quad (9.8)$$

$$BH_4(b+x) = AH_4(b)BH_4(x) + BH_4(b)AH_4(x)$$
$$+CH_4(b)DH_4(x) + DH_4(b)CH_4(x)$$
$$CH_4(b+x) = AH_4(b)CH_4(x) + BH_4(b)BH_4(x)$$
$$+CH_4(b)AH_4(x) + DH_4(b)DH_4(x) \quad (9.9)$$
$$DH_4(b+x) = AH_4(b)DH_4(x) + BH_4(b)CH_4(x)$$
$$+CH_4(b)BH_4(x) + DH_4(b)AH_4(x)$$

We have:

$$\det\left(\begin{bmatrix} w & x & y & z \\ z & w & x & y \\ y & z & w & x \\ x & y & z & w \end{bmatrix}\right) =$$

$$\det\left(PROD \begin{pmatrix} \begin{bmatrix} r & 0 & 0 & 0 \\ 0 & r & 0 & 0 \\ 0 & 0 & r & 0 \\ 0 & 0 & 0 & r \end{bmatrix} \\ \begin{bmatrix} AH_4(b) & BH_4(b) & CH_4(b) & DH_4(b) \\ DH_4(b) & AH_4(b) & BH_4(b) & CH_4(b) \\ CH_4(b) & DH_4(b) & AH_4(b) & BH_4(b) \\ BH_4(b) & CH_4(b) & DH_4(b) & AH_4(b) \end{bmatrix} \\ \begin{bmatrix} \cosh(c) & 0 & \sinh(c) & 0 \\ 0 & \cosh(c) & 0 & \sinh(c) \\ \sinh(c) & 0 & \cosh(c) & 0 \\ 0 & \sinh(c) & 0 & \cosh(c) \end{bmatrix} \\ \begin{bmatrix} AH_4(d) & DH_4(d) & CH_4(d) & BH_4(d) \\ BH_4(d) & AH_4(d) & DH_4(d) & CH_4(d) \\ CH_4(d) & BH_4(d) & AH_4(d) & DH_4(d) \\ DH_4(d) & CH_4(d) & BH_4(d) & AH_4(d) \end{bmatrix} \end{pmatrix}\right) = r^4 \quad (9.10)$$

Leading to the distance function:

$$r = \sqrt[4]{\begin{array}{l} a^4 - b^4 + c^4 - d^4 - 2a^2c^2 + 2b^2d^2 \\ -4a^2bd + 4ab^2c - 4bc^2d + 4acd^2 \end{array}} \quad (9.11)$$

The 4-dimensional simple-trig functions have representation as standard functions:

$$AH_4(x) = \frac{1}{4}(2\cosh(x) + 2\cos(x))$$

$$BH_4(x) = \frac{1}{4}(2\sinh(x) + 2\sin(x))$$

$$CH_4(x) = \frac{1}{4}(2\cosh(x) - 2\cos(x)) \quad (9.12)$$

$$DH_4(x) = \frac{1}{4}(2\sinh(x) - 2\sin(x))$$

The reader might be surprised by the occurrence of the 2-dimensional hyperbolic trigonometric functions within a matrix in the polar form of this 4-dimensional algebra. Such a matrix occurs in the polar form of every $\{C_{4.1}, C_{4.2}, C_{4.3}\}$ algebra. The polar form of the $C_2 \times C_2$ algebras has the form of three such matrices containing 2-dimensional trigonometric functions. The presence of these matrices is directly associated with the fact that the $\{C_{4.1}, C_{4.2}, C_{4.3}\}$ algebras have one 2-dimensional sub-algebra and the $C_2 \times C_2$ algebra has three 2-dimensional sub-algebras. We will consider more of this later.

The Polar Form of the $C_{4.1} L^1 E^3_{(j=-1, k=1, l=1)}$ Algebra:

The general form of the $C_{4.1}$ algebra is:

$$C_{4.1} = \begin{bmatrix} a & b & c & d \\ jd & a & kb & lc \\ \dfrac{jl}{k}c & \dfrac{j}{k}d & a & lb \\ jb & \dfrac{j}{k}c & \dfrac{j}{l}d & a \end{bmatrix} \quad (9.13)$$

We take $\{j = -1, k = 1, l = 1\}$ and exponentiate this matrix:

$$\exp\left(\begin{bmatrix} a & b & c & d \\ -d & a & b & c \\ -c & -d & a & b \\ -b & -c & -d & a \end{bmatrix}\right) = \quad (9.14)$$

$$PROD \begin{pmatrix} \begin{bmatrix} e^a & 0 & 0 & 0 \\ 0 & e^a & 0 & 0 \\ 0 & 0 & e^a & 0 \\ 0 & 0 & 0 & e^a \end{bmatrix} \\ \begin{bmatrix} AE_4(b) & BE_4(b) & CE_4(b) & DE_4(b) \\ -DE_4(b) & AE_4(b) & BE_4(b) & CE_4(b) \\ -CE_4(b) & -DE_4(b) & AE_4(b) & BE_4(b) \\ -BE_4(b) & -CE_4(b) & -DE_4(b) & AE_4(b) \end{bmatrix} \\ \begin{bmatrix} \cos(c) & 0 & \sin(c) & 0 \\ 0 & \cos(c) & 0 & \sin(c) \\ -\sin(c) & 0 & \cos(c) & 0 \\ 0 & -\sin(c) & 0 & \cos(c) \end{bmatrix} \\ \begin{bmatrix} AE_4(d) & DE_4(d) & -CE_4(d) & BE_4(d) \\ -BE_4(d) & AE_4(d) & DE_4(d) & -CE_4(d) \\ CE_4(d) & -BE_4(d) & AE_4(d) & DE_4(d) \\ -DE_4(d) & CE_4(d) & -BE_4(d) & AE_4(d) \end{bmatrix} \end{pmatrix} \quad (9.15)$$

Where:

$$\begin{aligned} AE_4(x) &= 1 - \frac{x^4}{4!} + \frac{x^8}{8!} - \ldots \\ BE_4(x) &= x - \frac{x^5}{5!} + \frac{x^9}{9!} - \ldots \\ CE_4(x) &= \frac{x^2}{2!} - \frac{x^6}{6!} + \frac{x^{10}}{10!} - \ldots \\ DE_4(x) &= \frac{x^3}{3!} - \frac{x^7}{7!} + \frac{x^{11}}{11!} - \ldots \end{aligned} \quad (9.16)$$

are the 4-way splittings of the exponential series (with some minus signs chucked in) that are the 4-dimensional E-type simple-trig functions.

Section 9

The matrices containing the simple-trig functions were formed by exponentiating a matrix with trace zero and thus the determinants of these matrices are unity. Thereby do we have a trigonometric identity.

$$\begin{aligned}
&AE_4^4(x) + BE_4^4(x) + CE_4^4(x) + DE_4^4(x) + 2AE_4^2(x)CE_4^2(x) \\
&+ 2BE_4^2(x)DE_4^2(x) + 4AE_4^2(x)BE_4(x)DE_4(x) \\
&- 4AE_4(x)BE_4^2(x)CE_4(x) - 4BE_4(x)CE_4^2(x)DE_4(x) \\
&+ 4AE_4(x)CE_4(x)DE_4^2(x) = 1
\end{aligned} \quad (9.17)$$

We also have the trigonometric identities:

$$\begin{aligned}
AE_4(b+x) &= AE_4(b)AE_4(x) - BE_4(b)DE_4(x) \\
&\quad - DE_4(b)BE_4(x) - CE_4(b)CE_4(x) \\
BE_4(b+x) &= AE_4(b)BE_4(x) + BE_4(b)AE_4(x) \\
&\quad - CE_4(b)DE_4(x) - DE_4(b)CE_4(x) \\
CE_4(b+x) &= AE_4(b)CE_4(x) + BE_4(b)BE_4(x) \\
&\quad + CE_4(b)AE_4(x) - DE_4(b)DE_4(x) \\
DE_4(b+x) &= AE_4(b)DE_4(x) + BE_4(b)CE_4(x) \\
&\quad + CE_4(b)BE_4(x) + DE_4(b)AE_4(x)
\end{aligned} \quad (9.18)$$

with a rather pretty arrangement of the minus signs.

We have the distance function:

$$r^4 = \det\left(C_{4,1} L^1 E^3_{(j=-1, k=1, l=1)}\right)$$

$$r = \sqrt[4]{\begin{array}{l} a^4 + b^4 + c^4 + d^4 + 2a^2c^2 + 2b^2d^2 \\ + 4a^2bd - 4ab^2c - 4bc^2d + 4acd^2 \end{array}} \quad (9.19)$$

The 4-dimensional simple-functions have representation as standard functions:

$$AE_4(x) = \cos\left(\frac{x}{\sqrt{2}}\right)\cosh\left(\frac{x}{\sqrt{2}}\right)$$

$$BE_4(x) = \frac{1}{\sqrt{2}}\left(\cos\left(\frac{x}{\sqrt{2}}\right)\sinh\left(\frac{x}{\sqrt{2}}\right) + \sin\left(\frac{x}{\sqrt{2}}\right)\cosh\left(\frac{x}{\sqrt{2}}\right)\right) \quad (9.20)$$

$$CE_4(x) = \sin\left(\frac{x}{\sqrt{2}}\right)\sinh\left(\frac{x}{\sqrt{2}}\right)$$

$$DE_4(x) = \frac{1}{\sqrt{2}}\left(-\cos\left(\frac{x}{\sqrt{2}}\right)\sinh\left(\frac{x}{\sqrt{2}}\right) + \sin\left(\frac{x}{\sqrt{2}}\right)\cosh\left(\frac{x}{\sqrt{2}}\right)\right) \quad (9.21)$$

The Polar Forms of the $\{C_{4.1}, C_{4.2}, C_{4.3}\}$ Algebras in General:

We have previously shown that the

$$\{C_{4.1} L^1 H^3{}_{(j=1,k=1,l=1)}, C_{4.2} L^1 H^3{}_{(j=1,k=1,l=1)}, C_{4.3} L^1 H^3{}_{(j=1,k=1,l=1)}\}$$

algebras are algebraically isomorphic to each other.

The reader is directed to inspect the appendices 5.1, 5.2, & 5.3 listing the polar forms of the flat $\{C_{4.1}, C_{4.2}, C_{4.3}\}$ algebras.

The eight flat polar forms of the $C_{4.1}$ algebra all have the same basic polar form. Other than the type (E-type or H-type) of simple-trig functions in the simple polar form matrices, they differ only in the way that the minus signs are scattered about. The same is true of the $\{C_{4.2}, C_{4.3}\}$ algebras. In two of the simple polar forms of the 3-dimensional flat algebras, we found both H-type and E-type simple-trig functions (though not in the same matrix). In the 4-dimensional natural algebras $\{C_{4.1}, C_{4.2}, C_{4.3}\}$, such mixing never occurs.[102]

The occurrence of a matrix containing the 2-dimensional trigonometric functions is connected to the existence of a 2-dimensional sub-algebra within the $\{C_{4.1}, C_{4.2}, C_{4.3}\}$ algebras. In the case of the $C_{4.1}$ algebras, this sub-algebra is formed from the $\{a,c\}$ variables:

$$\begin{bmatrix} a & 0 & c & 0 \\ 0 & a & 0 & lc \\ \frac{jl}{k}c & 0 & a & 0 \\ 0 & \frac{j}{k}c & 0 & a \end{bmatrix} \begin{bmatrix} e & 0 & f & 0 \\ 0 & e & 0 & lf \\ \frac{jl}{k}f & 0 & e & 0 \\ 0 & \frac{j}{k}f & 0 & e \end{bmatrix} = \quad (9.22)$$

---

[102] It does occur in some of the polar forms of the $C_2 \times C_2$ algebra.

$$\begin{bmatrix} ae + \dfrac{jl}{k}cf & 0 & af+ce & 0 \\ 0 & ae+\dfrac{jl}{k}cf & 0 & l(af+ce) \\ \dfrac{jl}{k}(af+ce) & 0 & ae+\dfrac{jl}{k}cf & 0 \\ 0 & \dfrac{j}{k}(af+ce) & 0 & ae+\dfrac{jl}{k}cf \end{bmatrix} \quad (9.23)$$

In the case of the C_{42} algebras, this sub-algebra is formed from the $\{a,b\}$ variables, and in the case of the C_{43} algebras, this sub-algebra is formed from the $\{a,d\}$ variables.[103]

The Polar Forms of the $C_2 \times C_2$ Algebras:

The reader is directed to inspect the polar forms of these algebras in appendix 5.4. The general 4-dimensional algebraic matrix form of the $C_2 \times C_2$ algebras is:

$$\begin{bmatrix} a & b & c & d \\ jb & a & \dfrac{j}{l}d & lc \\ kc & \dfrac{k}{l}d & a & lb \\ \dfrac{jk}{l^2}d & \dfrac{k}{l}c & \dfrac{j}{l}b & a \end{bmatrix} : \{l\} \neq 0 \quad (9.24)$$

With $\{j=-1, k=-1, l=1\}$, the polar form is:

---

[103] Each of these three algebras is essentially the same as the cyclic group C_4 with the C_2 subgroup being generated by a different one of each of the other non-identity elements.

$$\exp\left(\begin{bmatrix} a & b & c & d \\ -b & a & -d & c \\ -c & -d & a & b \\ d & -c & -b & a \end{bmatrix}\right) = \qquad (9.25)$$

$$PROD\left(\begin{bmatrix} h & 0 & 0 & 0 \\ 0 & h & 0 & 0 \\ 0 & 0 & h & 0 \\ 0 & 0 & 0 & h \end{bmatrix}\begin{bmatrix} \cos(b) & \sin(b) & 0 & 0 \\ -\sin(b) & \cos(b) & 0 & 0 \\ 0 & 0 & \cos(b) & \sin(b) \\ 0 & 0 & -\sin(b) & \cos(b) \end{bmatrix}\right.$$
$$\begin{bmatrix} \cos(c) & 0 & \sin(c) & 0 \\ 0 & \cos(c) & 0 & \sin(c) \\ -\sin(c) & 0 & \cos(c) & 0 \\ 0 & -\sin(c) & 0 & \cos(c) \end{bmatrix} \qquad (9.26)$$
$$\left.\begin{bmatrix} \cosh(d) & 0 & 0 & \sinh(d) \\ 0 & \cosh(d) & -\sinh(d) & 0 \\ 0 & -\sinh(d) & \cosh(d) & 0 \\ \sinh(d) & 0 & 0 & \cosh(d) \end{bmatrix}\right)$$

This polar form has three rotation matrices containing 2-dimensional trigonometric functions. All $C_2 \times C_2$ algebras similarly have a polar form with three such matrices. This contrasts with the cases of the other 4-dimensional algebraic matrix forms. In this case, as in most, there is a mixture of E-type trigonometric function matrices and H-type trigonometric function matrices.

The polar form reflects the fact that these algebras each have three 2-dimensional sub-algebras. In this case, we have:

Section 9

$$\begin{bmatrix} a & b & 0 & 0 \\ -b & a & 0 & 0 \\ 0 & 0 & a & b \\ 0 & 0 & -b & a \end{bmatrix} \begin{bmatrix} e & f & 0 & 0 \\ -f & e & 0 & 0 \\ 0 & 0 & e & f \\ 0 & 0 & -f & e \end{bmatrix} = \\ \begin{bmatrix} ae-bf & af+be & 0 & 0 \\ -(af+be) & ae-bf & 0 & 0 \\ 0 & 0 & ae-bf & af+be \\ 0 & 0 & -(af+be) & ae-bf \end{bmatrix}$$ (9.27)

$$\begin{bmatrix} a & 0 & c & 0 \\ 0 & a & 0 & c \\ -c & 0 & a & 0 \\ 0 & -c & 0 & a \end{bmatrix} \begin{bmatrix} e & 0 & g & 0 \\ 0 & e & 0 & g \\ -g & 0 & e & 0 \\ 0 & -g & 0 & e \end{bmatrix} = \\ \begin{bmatrix} ae-cg & 0 & ag+ce & 0 \\ 0 & ae-cg & 0 & ag+ce \\ -(ag+ce) & 0 & ae-cg & 0 \\ 0 & -(ag+ce) & 0 & ae-cg \end{bmatrix}$$ (9.28)

$$\begin{bmatrix} a & 0 & 0 & d \\ 0 & a & -d & 0 \\ 0 & -d & a & 0 \\ d & 0 & 0 & a \end{bmatrix} \begin{bmatrix} e & 0 & 0 & h \\ 0 & e & -h & 0 \\ 0 & -h & e & 0 \\ h & 0 & 0 & e \end{bmatrix} = \\ \begin{bmatrix} ae+dh & 0 & 0 & ah+de \\ 0 & ae+dh & -(ah+de) & 0 \\ 0 & -(ah+de) & ae+dh & 0 \\ ah+de & 0 & 0 & ae+dh \end{bmatrix}$$ (9.29)

The reader should compare these sub-algebras to the types of trigonometric functions within the polar form. There is a correspondence between the sub-algebras formed by the a variable and a particular variable and the type of trigonometric functions within the rotation matrix of that particular variable.

The distance function of this particular flat algebra is:

$$r = \sqrt[4]{\left((a-d)^2+(b+c)^2\right)\left((a+d)^2+(b-c)^2\right)} \qquad (9.30)$$

The General Polar Form $C_{4.1}$:

Taking the general polar form of the $C_{4.1}$ algebra leads to:

$$AH_4\left((-1)^{\frac{1}{4}}b\right) = AE_4(b) \qquad (9.31)$$

$$(-1)^{\frac{1}{4}}BH_4\left((-1)^{\frac{1}{4}}b\right) = BE_4(b)$$

$$(-1)^{\frac{2}{4}}CH_4\left((-1)^{\frac{1}{4}}b\right) = CE_4(b) \qquad (9.32)$$

$$(-1)^{\frac{3}{4}}DH_4\left((-1)^{\frac{1}{4}}b\right) = DE_4(b)$$

Quite Amazing!

Non-matrix Notation - $C_{4.1}L^1H^3{}_{(j=1,k=1,l=1)}$:

We have:

$$C_{4.1}L^1H^3{}_{(j=1,k=1,l=1)} = \begin{bmatrix} a & b & c & d \\ d & a & b & c \\ c & d & a & b \\ b & c & d & a \end{bmatrix} \qquad (9.33)$$

The non-matrix representation of this algebra is: $a + b\hat{r} + c\hat{s} + d\hat{t}$. The $\hat{r}$ is not the same object as we encountered in the hyperbolic complex numbers. It is just that we are running out of letters in the alphabet. Addition is component-wise:

$$a + b\hat{r} + c\hat{s} + d\hat{t} + e + f\hat{r} + g\hat{s} + h\hat{t} = (a+e) + (b+f)\hat{r} + (c+g)\hat{s} + (d+h)\hat{t}$$
$$(9.34)$$

Multiplication is taken (carefully) from the matrix product:

Section 9

$$\begin{bmatrix} a & b & c & d \\ d & a & b & c \\ c & d & a & b \\ b & c & d & a \end{bmatrix} \begin{bmatrix} e & f & g & h \\ h & e & f & g \\ g & h & e & f \\ f & g & h & e \end{bmatrix} = \begin{bmatrix} ae+bh+cg+df & \sim & \sim & \sim \\ ah+bg+cf+de & \sim & \sim & \sim \\ ag+bf+ce+dh & \sim & \sim & \sim \\ af+be+ch+dg & \sim & \sim & \sim \end{bmatrix} \quad (9.35)$$

$$\left(a+b\hat{r}+c\hat{s}+d\hat{t}\right)\left(e+f\hat{r}+g\hat{s}+h\hat{t}\right)$$
$$= (ae+bh+cg+df) + (af+be+ch+dg)\hat{r} \quad (9.36)$$
$$+ (ag+bf+ce+dh)\hat{s} + (ah+bg+cf+de)\hat{t}$$

The conjugate is taken from the adjoint matrix:

$$\mathrm{Adj}\left(\begin{bmatrix} a & b & c & d \\ d & a & b & c \\ c & d & a & b \\ b & c & d & a \end{bmatrix}\right) = \begin{bmatrix} a^3+cb^2-ac^2+cd^2-2abd & \sim & \sim & \sim \\ -b^3-a^2d-c^2d+bd^2+2abc & \sim & \sim & \sim \\ c^3-a^2c+ab^2+ad^2-2bcd & \sim & \sim & \sim \\ -d^3-a^2b+b^2d-bc^2+2acd & \sim & \sim & \sim \end{bmatrix} \quad (9.37)$$

$$\left(a+b\hat{r}+c\hat{s}+d\hat{t}\right) =$$
$$\left(a^3+cb^2-ac^2+cd^2-2abd\right) + \left(-d^3-a^2b+b^2d-bc^2+2acd\right)\hat{r}$$
$$+ \left(c^3-a^2c+ab^2+ad^2-2bcd\right)\hat{s} + \left(-b^3-a^2d-c^2d+bd^2+2abc\right)\hat{t}$$
$$(9.38)$$

Term by term expansion of the multiplication operation gives:

$$\left(a+b\hat{r}+c\hat{s}+d\hat{t}\right)\left(e+f\hat{r}+g\hat{s}+h\hat{t}\right) =$$
$$ae + \hat{r}(af+be) + \hat{s}(ag+ce) + \hat{t}(ah+de)$$
$$+ (bg+cf)\hat{r}\hat{s} + (bh+df)\hat{r}\hat{t} + (ch+dg)\hat{s}\hat{t} \quad (9.39)$$
$$+ bf\hat{r}^2 + cg\hat{s}^2 + dh\hat{t}^2 =$$
$$(ae+bh+cg+df) + (af+be+ch+dg)\hat{r}$$
$$+ (ag+bf+ce+dh)\hat{s} + (ah+bg+cf+de)\hat{t}$$

Inspection of this leads to:

$$\left\{ \begin{aligned} \hat{r}^2 = \hat{s}, \hat{s}^2 = 1, \hat{t}^2 = \hat{s}, \hat{r}\hat{s} = \hat{t}, \hat{r}\hat{t} = 1, \hat{s}\hat{t} = \hat{r} \\ \hat{s} = \sqrt[2]{+1}, \hat{r} = \sqrt[4]{+1}, \hat{t} = \sqrt[4]{+1} \end{aligned} \right\} \quad (9.40)$$

The form: $a+c\hat{s}$ is the hyperbolic complex number sub-algebra. We have:

$$e^{\hat{r}\theta} = 1 + \hat{r}\theta + \frac{\hat{r}^2\theta^2}{2!} + \frac{\hat{r}^3\theta^3}{3!} + \frac{\hat{r}^4\theta^4}{4!} + \frac{\hat{r}^5\theta^5}{5!} + \frac{\hat{r}^6\theta^6}{6!} + \ldots$$

$$= 1 + \hat{r}\theta + \frac{\hat{s}\theta^2}{2!} + \frac{\hat{t}\theta^3}{3!} + \frac{\theta^4}{4!} + \frac{\hat{r}\theta^5}{5!} + \frac{\hat{s}\theta^6}{6!} + \ldots$$

$$= \left(1 + \frac{\theta^4}{4!} + \ldots\right) + \hat{r}\left(\theta + \frac{\theta^5}{5!} + \ldots\right) + \hat{s}\left(\frac{\theta^2}{2!} + \frac{\theta^6}{6!} + \ldots\right) + \hat{t}\left(\frac{\theta^3}{3!} + \frac{\theta^7}{7!} + \ldots\right)$$

$$= AH_4(\theta) + \hat{r}BH_4(\theta) + \hat{s}CH_4(\theta) + \hat{t}DH_4(\theta)$$

(9.41)

And:

$$e^{\hat{s}\theta} = 1 + \hat{s}\theta + \frac{\hat{s}^2\theta^2}{2!} + \frac{\hat{s}^3\theta^3}{3!} + \frac{\hat{s}^4\theta^4}{4!} + \frac{\hat{s}^5\theta^5}{5!} + \frac{\hat{s}^6\theta^6}{6!} + \ldots$$

$$= 1 + \hat{s}\theta + \frac{\theta^2}{2!} + \frac{\hat{s}\theta^3}{3!} + \frac{\theta^4}{4!} + \frac{\hat{s}\theta^5}{5!} + \frac{\theta^6}{6!} + \ldots \qquad (9.42)$$

$$= \left(1 + \frac{\theta^2}{2!} + \frac{\theta^4}{4!} + \frac{\theta^6}{6!} + \ldots\right) + \hat{s}\left(\theta + \frac{\theta^3}{3!} + \frac{\theta^5}{5!} + \ldots\right)$$

$$= \cos(\theta) + \hat{s}\sin(\theta)$$

And:

$$e^{\hat{t}\theta} = 1 + \hat{t}\theta + \frac{\hat{t}^2\theta^2}{2!} + \frac{\hat{t}^3\theta^3}{3!} + \frac{\hat{t}^4\theta^4}{4!} + \frac{\hat{t}^5\theta^5}{5!} + \frac{\hat{t}^6\theta^6}{6!} + \ldots$$

$$= 1 + \hat{t}\theta + \frac{\hat{s}\theta^2}{2!} + \frac{\hat{r}\theta^3}{3!} + \frac{\theta^4}{4!} + \frac{\hat{t}\theta^5}{5!} + \frac{\hat{s}\theta^6}{6!} + \ldots$$

$$= \left(1 + \frac{\theta^4}{4!} + \ldots\right) + \hat{t}\left(\theta + \frac{\theta^5}{5!} + \ldots\right) + \hat{s}\left(\frac{\theta^2}{2!} + \frac{\theta^6}{6!} + \ldots\right) + \hat{r}\left(\frac{\theta^3}{3!} + \ldots\right)$$

$$= AH_4(\theta) + \hat{t}BH_4(\theta) + \hat{s}CH_4(\theta) + \hat{r}DH_4(\theta)$$

(9.43)

Leading to:

Section 9

$$e^{b\hat{r}+c\hat{s}+d\hat{t}} = e^{b\hat{r}}e^{c\hat{s}}e^{d\hat{t}}$$

$$= PROD \begin{pmatrix} AH_4(b)+\hat{r}AH_4(b)+\hat{s}AH_4(b)+\hat{t}AH_4(b) \\ \cos(c)+\hat{s}\sin(c) \\ AH_4(d)+\hat{t}AH_4(d)+\hat{s}AH_4(d)+\hat{r}AH_4(d) \end{pmatrix}$$

$$= v_{[\,]}A(b,c,d)+\hat{r}v_{[\,]}B(b,c,d)+\hat{s}v_{[\,]}C(b,c,d)+\hat{t}v_{[\,]}D(b,c,d)$$

(9.44)

And:

$$e^{\frac{b\hat{r}+c\hat{s}+d\hat{t}}{n}} = v_{[\,]}A\left(\frac{b}{n},\frac{c}{n},\frac{d}{n}\right)+\hat{r}v_{[\,]}B\left(\frac{b}{n},\frac{c}{n},\frac{d}{n}\right)$$
$$+\hat{s}v_{[\,]}C\left(\frac{b}{n},\frac{c}{n},\frac{d}{n}\right)+\hat{t}v_{[\,]}D\left(\frac{b}{n},\frac{c}{n},\frac{d}{n}\right)$$

(9.45)

The Euclidean shadow of the $C_{4,1}L^1H^3_{(j=1,k=1,l=1)}$ Algebra:

In non-matrix notation, the $C_{4,1}L^1H^3_{(j=1,k=1,l=1)}$ algebra is the algebra of $a+b\hat{r}+c\hat{s}+d\hat{t}$. The multiplicative relations are:

$$\{\hat{r}\hat{s}=\hat{t}, \hat{r}\hat{t}=1, \hat{s}\hat{t}=\hat{r}, \hat{r}^2=\hat{s}, \hat{s}^2=1, \hat{t}^2=\hat{s}\} : \hat{r}\neq\hat{s}\neq\hat{t}$$ (9.46)

The complex 4th roots of unity have the same multiplicative relations with:

$$\{\hat{r}=\hat{i}, \hat{s}=-1, \hat{t}=-\hat{i}\}$$ (9.47)

Substituting these complex roots into the non-matrix representation of the $C_{4,1}L^1H^3_{(j=1,k=1,l=1)}$ algebra, we get:

$$a+b\hat{r}+c\hat{s}+d\hat{t} = (a-c)+\hat{i}(b-d)$$ (9.48)

The polar form of this representation is:

$$\exp\left(\begin{bmatrix} a-c & b-d \\ -(b-d) & a-c \end{bmatrix}\right) = \begin{bmatrix} \frac{e^a}{e^c} & 0 \\ 0 & \frac{e^a}{e^c} \end{bmatrix} \begin{bmatrix} \cos(b-d) & \sin(b-d) \\ -\sin(b-d) & \cos(b-d) \end{bmatrix}$$

(9.49)

We see that the length within the 4-dimensional algebra, e^a, has now become $\frac{e^a}{e^c}$, and the 4-dimensional $\{b,c,d\}$ rotation matrices have been combined into a single 2-dimensional rotation matrix, but that the c rotation matrix is not represented in this rotation matrix.

The determinant of this complex number is:

$$(a-c)^2 + (b-d)^2 \tag{9.50}$$

Which the reader should compare to the determinant of the $C_{4,1}L^1H^3_{(j=1,k=1,l=1)}$ algebra which is:

$$\left((a-c)^2 + (b-d)^2\right)\left((a+c)^2 - (b+d)^2\right) \tag{9.51}$$

The factor $(a+c)^2 - (b+d)^2$ is of the form of the 2-dimensional hyperbolic metric – we seem to have squashed out 2 hyperbolic dimensions.

The Cauchy Riemann Equations - $C_{4,2}L^1H^3_{(j=1,k=1,l=1)}$:

We have:

$$\begin{bmatrix} t & x & y & z \\ x & t & z & y \\ z & y & t & x \\ y & z & x & t \end{bmatrix} \rightarrow \begin{bmatrix} s(t,x,y,z) & f(t,x,y,z) & g(t,x,y,z) & h(t,x,y,z) \\ f(t,x,y,z) & s(t,x,y,z) & h(t,x,y,z) & g(t,x,y,z) \\ h(t,x,y,z) & g(t,x,y,z) & s(t,x,y,z) & f(t,x,y,z) \\ g(t,x,y,z) & h(t,x,y,z) & f(t,x,y,z) & s(t,x,y,z) \end{bmatrix} \tag{9.52}$$

$$\begin{bmatrix} s & f & g & h \\ f & s & h & g \\ h & g & s & f \\ g & h & f & s \end{bmatrix} = \tag{9.53}$$

Section 9

$$\begin{bmatrix} 1 & 0 & 0 & 0 \\ 0 & 1 & 0 & 0 \\ 0 & 0 & 1 & 0 \\ 0 & 0 & 0 & 1 \end{bmatrix} \begin{bmatrix} s & 0 & 0 & 0 \\ 0 & s & 0 & 0 \\ 0 & 0 & s & 0 \\ 0 & 0 & 0 & s \end{bmatrix} + \begin{bmatrix} 0 & 1 & 0 & 0 \\ 1 & 0 & 0 & 0 \\ 0 & 0 & 0 & 1 \\ 0 & 0 & 1 & 0 \end{bmatrix} \begin{bmatrix} f & 0 & 0 & 0 \\ 0 & f & 0 & 0 \\ 0 & 0 & f & 0 \\ 0 & 0 & 0 & f \end{bmatrix}$$
$$+ \begin{bmatrix} 0 & 0 & 1 & 0 \\ 0 & 0 & 0 & 1 \\ 0 & 1 & 0 & 0 \\ 1 & 0 & 0 & 0 \end{bmatrix} \begin{bmatrix} g & 0 & 0 & 0 \\ 0 & g & 0 & 0 \\ 0 & 0 & g & 0 \\ 0 & 0 & 0 & g \end{bmatrix} + \begin{bmatrix} 0 & 0 & 0 & 1 \\ 0 & 0 & 1 & 0 \\ 1 & 0 & 0 & 0 \\ 0 & 1 & 0 & 0 \end{bmatrix} \begin{bmatrix} h & 0 & 0 & 0 \\ 0 & h & 0 & 0 \\ 0 & 0 & h & 0 \\ 0 & 0 & 0 & h \end{bmatrix} \quad (9.54)$$

Differentiating with respect to:

$$\begin{bmatrix} t & 0 & 0 & 0 \\ 0 & t & 0 & 0 \\ 0 & 0 & t & 0 \\ 0 & 0 & 0 & t \end{bmatrix} \text{ leads to } \begin{bmatrix} \dfrac{\partial s}{\partial t} & \dfrac{\partial f}{\partial t} & \dfrac{\partial g}{\partial t} & \dfrac{\partial h}{\partial t} \\ \dfrac{\partial f}{\partial t} & \dfrac{\partial s}{\partial t} & \dfrac{\partial h}{\partial t} & \dfrac{\partial g}{\partial t} \\ \dfrac{\partial h}{\partial t} & \dfrac{\partial g}{\partial t} & \dfrac{\partial s}{\partial t} & \dfrac{\partial f}{\partial t} \\ \dfrac{\partial g}{\partial t} & \dfrac{\partial h}{\partial t} & \dfrac{\partial f}{\partial t} & \dfrac{\partial s}{\partial t} \end{bmatrix} \quad (9.55)$$

Differentiating with respect to:

$$\begin{bmatrix} 0 & 1 & 0 & 0 \\ 1 & 0 & 0 & 0 \\ 0 & 0 & 0 & 1 \\ 0 & 0 & 1 & 0 \end{bmatrix} \begin{bmatrix} x & 0 & 0 & 0 \\ 0 & x & 0 & 0 \\ 0 & 0 & x & 0 \\ 0 & 0 & 0 & x \end{bmatrix} \text{ leads to } \begin{bmatrix} \dfrac{\partial f}{\partial x} & \dfrac{\partial s}{\partial x} & \dfrac{\partial h}{\partial x} & \dfrac{\partial g}{\partial x} \\ \dfrac{\partial s}{\partial x} & \dfrac{\partial f}{\partial x} & \dfrac{\partial g}{\partial x} & \dfrac{\partial h}{\partial x} \\ \dfrac{\partial g}{\partial x} & \dfrac{\partial h}{\partial x} & \dfrac{\partial f}{\partial x} & \dfrac{\partial s}{\partial x} \\ \dfrac{\partial h}{\partial x} & \dfrac{\partial g}{\partial x} & \dfrac{\partial s}{\partial x} & \dfrac{\partial f}{\partial x} \end{bmatrix} \quad (9.56)$$

Differentiating with respect to:

$$\begin{bmatrix} 0 & 0 & 1 & 0 \\ 0 & 0 & 0 & 1 \\ 0 & 1 & 0 & 0 \\ 1 & 0 & 0 & 0 \end{bmatrix} \begin{bmatrix} y & 0 & 0 & 0 \\ 0 & y & 0 & 0 \\ 0 & 0 & y & 0 \\ 0 & 0 & 0 & y \end{bmatrix} \text{ leads to } \begin{bmatrix} \dfrac{\partial g}{\partial y} & \dfrac{\partial h}{\partial y} & \dfrac{\partial f}{\partial y} & \dfrac{\partial s}{\partial y} \\ \dfrac{\partial h}{\partial y} & \dfrac{\partial g}{\partial y} & \dfrac{\partial s}{\partial y} & \dfrac{\partial f}{\partial y} \\ \dfrac{\partial s}{\partial y} & \dfrac{\partial f}{\partial y} & \dfrac{\partial g}{\partial y} & \dfrac{\partial h}{\partial y} \\ \dfrac{\partial f}{\partial y} & \dfrac{\partial s}{\partial y} & \dfrac{\partial h}{\partial y} & \dfrac{\partial g}{\partial y} \end{bmatrix} \quad (9.57)$$

Differentiating with respect to:

$$\begin{bmatrix} 0 & 0 & 0 & 1 \\ 0 & 0 & 1 & 0 \\ 1 & 0 & 0 & 0 \\ 0 & 1 & 0 & 0 \end{bmatrix} \begin{bmatrix} z & 0 & 0 & 0 \\ 0 & z & 0 & 0 \\ 0 & 0 & z & 0 \\ 0 & 0 & 0 & z \end{bmatrix} \text{ leads to } \begin{bmatrix} \dfrac{\partial h}{\partial z} & \dfrac{\partial g}{\partial z} & \dfrac{\partial s}{\partial z} & \dfrac{\partial f}{\partial z} \\ \dfrac{\partial g}{\partial z} & \dfrac{\partial h}{\partial z} & \dfrac{\partial f}{\partial z} & \dfrac{\partial s}{\partial z} \\ \dfrac{\partial f}{\partial z} & \dfrac{\partial s}{\partial z} & \dfrac{\partial h}{\partial z} & \dfrac{\partial g}{\partial z} \\ \dfrac{\partial s}{\partial z} & \dfrac{\partial f}{\partial z} & \dfrac{\partial g}{\partial z} & \dfrac{\partial h}{\partial z} \end{bmatrix} \quad (9.58)$$

This leads to the Cauchy Riemann equations of the $C_{4,2}L^1H^3_{(j=1,k=1,l=1)}$, which are:

$$\frac{\partial s}{\partial t} = \frac{\partial f}{\partial x} = \frac{\partial g}{\partial y} = \frac{\partial h}{\partial z}$$

$$\frac{\partial f}{\partial t} = \frac{\partial s}{\partial x} = \frac{\partial h}{\partial y} = \frac{\partial g}{\partial z}$$

$$\frac{\partial g}{\partial t} = \frac{\partial h}{\partial x} = \frac{\partial f}{\partial y} = \frac{\partial s}{\partial z}$$

$$\frac{\partial h}{\partial t} = \frac{\partial g}{\partial x} = \frac{\partial s}{\partial y} = \frac{\partial f}{\partial z}$$

(9.59)

Section 10

Introduction to Section 10:

In this section, we look more closely at the 4-dimensional simple-trig functions and some of the 4-dimensional v-functions. The reader is pointed toward the graphs of the simple-trig functions. This section contains no great insights, but it will familiarise the reader with 4-dimensional trigonometry.

The 4-dimensional Simple-trig Functions:

The 4-dimensional simple-trig functions are defined as 4-way splittings of the exponential series.

$$AH_4(x) = hypergeom\left([\ ], \left[\frac{1}{4}, \frac{1}{2}, \frac{3}{4}\right], \left(\frac{x}{4}\right)^4\right) = 1 + \frac{x^4}{4!} + \frac{x^8}{8!} + \frac{x^{12}}{12!} + \ldots$$

$$BH_4(x) = x\left(hypergeom\left([\ ], \left[\frac{1}{2}, \frac{3}{4}, \frac{5}{4}\right], \left(\frac{x}{4}\right)^4\right)\right) = x + \frac{x^5}{5!} + \frac{x^9}{9!} + \frac{x^{13}}{13!} + \ldots$$

$$CH_4(x) = \frac{x^2}{2}\left(hypergeom\left([\ ], \left[\frac{3}{4}, \frac{5}{4}, \frac{6}{4}\right], \left(\frac{x}{4}\right)^4\right)\right) = \frac{x^2}{2!} + \frac{x^6}{6!} + \frac{x^{10}}{10!} + \ldots$$

$$DH_4(x) = \frac{x^3}{6}\left(hypergeom\left([\ ], \left[\frac{5}{4}, \frac{6}{4}, \frac{7}{4}\right], \left(\frac{x}{4}\right)^4\right)\right) = \frac{x^3}{3!} + \frac{x^7}{7!} + \frac{x^{11}}{11!} + \ldots$$

(10.1)

$$AE_4(x) = hypergeom\left([\ ], \left[\frac{1}{4}, \frac{1}{2}, \frac{3}{4}\right], -\left(\frac{x}{4}\right)^4\right) = 1 - \frac{x^4}{4!} + \frac{x^8}{8!} - \frac{x^{12}}{12!} + \ldots$$

$$BE_4(x) = x\left(hypergeom\left([\ ], \left[\frac{1}{2}, \frac{3}{4}, \frac{5}{4}\right], -\left(\frac{x}{4}\right)^4\right)\right) = x - \frac{x^5}{5!} + \frac{x^9}{9!} - \frac{x^{13}}{13!} + \ldots$$

$$CE_4(x) = \frac{x^2}{2}\left(hypergeom\left([\ ], \left[\frac{3}{4}, \frac{5}{4}, \frac{6}{4}\right], -\left(\frac{x}{4}\right)^4\right)\right) = \frac{x^2}{2!} - \frac{x^6}{6!} + \frac{x^{10}}{10!} - \ldots$$

$$DE_4(x) = \frac{x^3}{6}\left(hypergeom\left([\ ], \left[\frac{5}{4}, \frac{6}{4}, \frac{7}{4}\right], -\left(\frac{x}{4}\right)^4\right)\right) = \frac{x^3}{3!} - \frac{x^7}{7!} + \frac{x^{11}}{11!} - \ldots$$

(10.2)

Clearly:

Complex Numbers The Higher Dimensional Forms – 2nd Edition

$$AH_4(x) + BH_4(x) + CH_4(x) + DH_4(x) = e^x$$
$$AH_4(x) + CH_4(x) = \cosh(x)$$
$$BH_4(x) + DH_4(x) = \sinh(x) \quad (10.3)$$
$$AH_4(x) - CH_4(x) = \cos(x)$$
$$BH_4(x) - DH_4(x) = \sin(x)$$

As with all simple-trig functions, these have a differentiation cycle.[104] We have:

$$\frac{d}{dx} AH_4(x) = DH_4(x) \qquad \frac{d}{dx} AE_4(x) = -DE_4(x)$$
$$\frac{d}{dx} BH_4(x) = AH_4(x) \qquad \frac{d}{dx} BE_4(x) = AE_4(x)$$
$$\frac{d}{dx} CH_4(x) = BH_4(x) \qquad \frac{d}{dx} CE_4(x) = BE_4(x) \quad (10.4)$$
$$\frac{d}{dx} DH_4(x) = CH_4(x) \qquad \frac{d}{dx} DE_4(x) = CE_4(x)$$

In standard form, the 4-dimensional H-type simple-trig functions are:

$$AH_4(x) = \frac{1}{4}(2\cosh x + 2\cos x)$$
$$BH_4(x) = \frac{1}{4}(2\sinh x + 2\sin x)$$
$$CH_4(x) = \frac{1}{4}(2\cosh x - 2\cos x) \quad (10.5)$$
$$DH_4(x) = \frac{1}{4}(2\sinh x - 2\sin x)$$

When $x = 0$:

$$AH_4(0) = 1$$
$$BH_4(0) = 0$$
$$CH_4(0) = 0 \quad (10.6)$$
$$DH_4(0) = 0$$

---

[104] As is clear from examination of the series expansions.

Section 10

This phenomenon of the A simple-trig function being unity when the argument is zero and all the other simple-trig functions being zero when the argument is zero is general to all dimensions.[105]

The graphs of the 4-dimensional H-type simple-trig functions are:

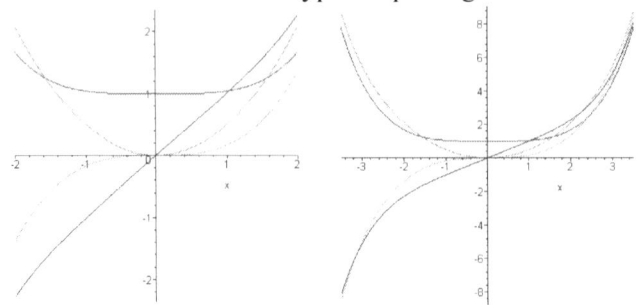

Which is reminiscent of the graphs of *cosh* and *sinh*. The exponential term ensures that, for large positive x, these three functions are effectively equal. These functions are not periodic.

The 4-dimensional E-type simple-trig functions are:

$$AE_4(x) = \cos\left(\frac{x}{\sqrt{2}}\right)\cosh\left(\frac{x}{\sqrt{2}}\right)$$

$$BE_4(x) = \frac{1}{\sqrt{2}}\left(\cos\left(\frac{x}{\sqrt{2}}\right)\sinh\left(\frac{x}{\sqrt{2}}\right) + \cosh\left(\frac{x}{\sqrt{2}}\right)\sin\left(\frac{x}{\sqrt{2}}\right)\right)$$

$$CE_4(x) = \sin\left(\frac{x}{\sqrt{2}}\right)\sinh\left(\frac{x}{\sqrt{2}}\right)$$

$$DE_4(x) = \frac{1}{\sqrt{2}}\left(\sin\left(\frac{x}{\sqrt{2}}\right)\cosh\left(\frac{x}{\sqrt{2}}\right) - \cos\left(\frac{x}{\sqrt{2}}\right)\sinh\left(\frac{x}{\sqrt{2}}\right)\right)$$

(10.7)

The graphs of the 4-dimensional E-type simple-trig functions are:

---

[105] Just look at the series expansions.

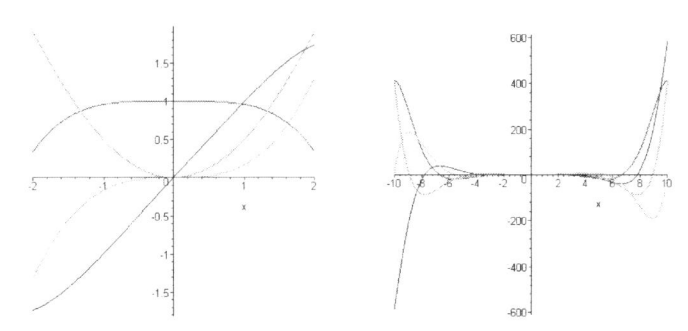

These functions are wave-like with enormous amplitude for large positive and large negative values of x; they are periodic.

The v-functions - $C_{4.1} L^1 H^3{}_{(j=1, k=1, l=1)}$:

The simple polar form of the $C_{4.1} L^1 H^3{}_{(j=1, k=1, l=1)}$ algebra is:

Section 10

$$\exp\left(\begin{bmatrix} a & b & c & d \\ d & a & b & c \\ c & d & a & b \\ b & c & d & a \end{bmatrix}\right) =$$

$$PROD \begin{pmatrix} \begin{bmatrix} e^a & 0 & 0 & 0 \\ 0 & e^a & 0 & 0 \\ 0 & 0 & e^a & 0 \\ 0 & 0 & 0 & e^a \end{bmatrix} \\ \begin{bmatrix} AH_4(b) & BH_4(b) & CH_4(b) & DH_4(b) \\ DH_4(b) & AH_4(b) & BH_4(b) & CH_4(b) \\ CH_4(b) & DH_4(b) & AH_4(b) & BH_4(b) \\ BH_4(b) & CH_4(b) & DH_4(b) & AH_4(b) \end{bmatrix} \\ \begin{bmatrix} \cosh(c) & 0 & \sinh(c) & 0 \\ 0 & \cosh(c) & 0 & \sinh(c) \\ \sinh(c) & 0 & \cosh(c) & 0 \\ 0 & \sinh(c) & 0 & \cosh(c) \end{bmatrix} \\ \begin{bmatrix} AH_4(d) & DH_4(d) & CH_4(d) & BH_4(d) \\ BH_4(d) & AH_4(d) & DH_4(d) & CH_4(d) \\ CH_4(d) & BH_4(d) & AH_4(d) & DH_4(d) \\ DH_4(d) & CH_4(d) & BH_4(d) & AH_4(d) \end{bmatrix} \end{pmatrix} \quad (10.8)$$

Multiplying the three rotation matrices together gives the compound rotation matrix containing the 4-dimensional v-functions of this algebra:

$$\begin{bmatrix} v_{[\,]}A(b,c,d) & v_{[\,]}B(b,c,d) & v_{[\,]}C(b,c,d) & v_{[\,]}D(b,c,d) \\ v_{[\,]}D(b,c,d) & v_{[\,]}A(b,c,d) & v_{[\,]}B(b,c,d) & v_{[\,]}C(b,c,d) \\ v_{[\,]}C(b,c,d) & v_{[\,]}D(b,c,d) & v_{[\,]}A(b,c,d) & v_{[\,]}B(b,c,d) \\ v_{[\,]}B(b,c,d) & v_{[\,]}C(b,c,d) & v_{[\,]}D(b,c,d) & v_{[\,]}A(b,c,d) \end{bmatrix} \quad (10.9)$$

We have:

$$v_{\left[C_n \dot{L}H^i_{(j=1,k=1,l=1)}\right]}A(b,c,d) = \frac{1}{4}\left(2e^c \cosh(b+d) + 2e^{-c}\cos(b-d)\right)$$

$$v_{\left[C_n \dot{L}H^i_{(j=1,k=1,l=1)}\right]}B(b,c,d) = \frac{1}{4}\left(2e^c \sinh(b+d) + 2e^{-c}\sin(b-d)\right)$$

$$v_{\left[C_n \dot{L}H^i_{(j=1,k=1,l=1)}\right]}C(b,c,d) = \frac{1}{4}\left(2e^c \cosh(b+d) - 2e^{-c}\cos(b-d)\right)$$

$$v_{\left[C_n \dot{L}H^i_{(j=1,k=1,l=1)}\right]}D(b,c,d) = \frac{1}{4}\left(2e^c \sinh(b+d) - 2e^{-c}\sin(b-d)\right)$$

(10.10)

We have:

$$v_{\left[C_{4,1}\dot{L}H^3_{(j=1,k=1,l=1)}\right]}A(0,0,d) = \frac{1}{4}\left(2\cosh(d) + 2\cos(d)\right)$$

$$v_{\left[C_{4,1}\dot{L}H^3_{(j=1,k=1,l=1)}\right]}A(0,c,0) = \cosh(c) \qquad (10.11)$$

$$v_{\left[C_{4,1}\dot{L}H^3_{(j=1,k=1,l=1)}\right]}A(b,0,0) = \frac{1}{4}\left(2\cosh(b) + 2\cos(b)\right)$$

And:

$$v_{\left[C_{4,1}\dot{L}H^3_{(j=1,k=1,l=1)}\right]}B(0,0,d) = \frac{1}{4}\left(2\sinh(d) - 2\sin(d)\right)$$

$$v_{\left[C_{4,1}\dot{L}H^3_{(j=1,k=1,l=1)}\right]}B(0,c,0) = 0 \qquad (10.12)$$

$$v_{\left[C_{4,1}\dot{L}H^3_{(j=1,k=1,l=1)}\right]}B(b,0,0) = \frac{1}{4}\left(2\sinh(b) + 2\sin(b)\right)$$

And:

$$v_{\left[C_{4,1}\dot{L}H^3_{(j=1,k=1,l=1)}\right]}C(0,0,d) = \frac{1}{4}\left(2\cosh(d) - 2\cos(d)\right)$$

$$v_{\left[C_{4,1}\dot{L}H^3_{(j=1,k=1,l=1)}\right]}C(0,c,0) = \sinh(c) \qquad (10.13)$$

$$v_{\left[C_{4,1}\dot{L}H^3_{(j=1,k=1,l=1)}\right]}C(b,0,0) = \frac{1}{4}\left(2\cosh(b) - 2\cos(b)\right)$$

And:

$$v_{\left[C_{4,1}\dot{L}H^3_{(j=1,k=1,l=1)}\right]}D(0,0,d) = \frac{1}{4}\left(2\sinh(d) + 2\sin(d)\right)$$

$$v_{\left[C_{4,1}\dot{L}H^3_{(j=1,k=1,l=1)}\right]}D(0,c,0) = 0 \qquad (10.14)$$

$$v_{\left[C_{4,1}\dot{L}H^3_{(j=1,k=1,l=1)}\right]}D(b,0,0) = \frac{1}{4}\left(2\sinh(b) - 2\sin(b)\right)$$

Section 10

We have:

$$v_{[\ ]}A(b,c,d) + v_{[\ ]}B(b,c,d) + v_{[\ ]}C(b,c,d) + v_{[\ ]}D(b,c,d) = e^{b+c+d}$$
(10.15)

The differentiation cycles are:

$$\frac{\partial(v_{[\ ]}A(b,c,d))}{\partial b} = v_{[\ ]}D(b,c,d)$$

$$\frac{\partial(v_{[\ ]}A(b,c,d))}{\partial c} = v_{[\ ]}C(b,c,d) \quad (10.16)$$

$$\frac{\partial(v_{[\ ]}A(b,c,d))}{\partial d} = v_{[\ ]}B(b,c,d)$$

$$\frac{\partial(v_{[\ ]}B(b,c,d))}{\partial b} = v_{[\ ]}A(b,c,d)$$

$$\frac{\partial(v_{[\ ]}B(b,c,d))}{\partial c} = v_{[\ ]}D(b,c,d) \quad (10.17)$$

$$\frac{\partial(v_{[\ ]}B(b,c,d))}{\partial d} = v_{[\ ]}C(b,c,d)$$

$$\frac{\partial(v_{[\ ]}C(b,c,d))}{\partial b} = v_{[\ ]}B(b,c,d)$$

$$\frac{\partial(v_{[\ ]}C(b,c,d))}{\partial c} = v_{[\ ]}A(b,c,d) \quad (10.18)$$

$$\frac{\partial(v_{[\ ]}C(b,c,d))}{\partial d} = v_{[\ ]}D(b,c,d)$$

$$\frac{\partial(v_{[\ ]}D(b,c,d))}{\partial b} = v_{[\ ]}C(b,c,d)$$

$$\frac{\partial(v_{[\ ]}D(b,c,d))}{\partial c} = v_{[\ ]}B(b,c,d) \quad (10.19)$$

$$\frac{\partial(v_{[\ ]}D(b,c,d))}{\partial d} = v_{[\ ]}A(b,c,d)$$

The $C_{4.1}L^1H^3_{(j=1,k=1,l=1)}$ v-functions have series expansions. We list the case of $v_{[\ ]}A$:

Expanded about $b = 0$:

$$v_{[\ ]}A(b,c,d) = v_{[\ ]}A(0,c,d) + v_{[\ ]}B(0,c,d)b + v_{[\ ]}C(0,c,d)\frac{b^2}{2!}$$

$$+ v_{[\ ]}D(0,c,d)\frac{b^3}{3!} + v_{[\ ]}A(0,c,d)\frac{b^4}{4!} + \ldots$$

(10.20)

This series is a sum of all four v-functions.

Expanded about $c = 0$:

$$v_{[\ ]}A(b,c,d) = v_{[\ ]}A(b,0,d) + v_{[\ ]}C(b,0,d)c + v_{[\ ]}A(b,0,d)\frac{c^2}{2!}$$

$$+ v_{[\ ]}C(b,0,d)\frac{c^3}{3!} + v_{[\ ]}A(b,0,d)\frac{c^4}{4!} + \ldots$$

(10.21)

This series is a sum of only two v-functions.

Expanded about $d = 0$:

$$v_{[\ ]}A(b,c,d) = v_{[\ ]}A(b,c,0) + v_{[\ ]}B(b,c,0)d + v_{[\ ]}C(b,c,0)\frac{d^2}{2!}$$

$$+ v_{[\ ]}D(b,c,0)\frac{d^3}{3!} + v_{[\ ]}A(b,c,0)\frac{d^4}{4!} + \ldots$$

(10.22)

This series is a sum of all four v-functions.

The series form of the 3-dimensional v-functions, is a sum of the 3-dimensional simple-trig functions. However, in 3-dimensions, the v-functions with one argument equal to zero are the simple-trig functions. Thus, we could equally well have written the 3-dimensional series with v-functions with one argument zero instead of the simple-trig functions. Since the v-functions have a differentiation cycle and the series are calculated as a sum of the differentials, this is not surprising.

Section 10
Addendum

The v-functions - $C_{4,2}L^1H^3{}_{(j=1,k=1,l=1)}$:

We have:

$$v_{[C_{4,2}L^1H^3{}_{(j-1,k-1,j-1)}]}A(b,c,d) = \frac{1}{4}\left(2e^b \cosh(c+d) + 2e^{-b}\cos(c-d)\right)$$

$$v_{[C_{4,2}L^1H^3{}_{(j-1,k-1,j-1)}]}B(b,c,d) = \frac{1}{4}\left(2e^b \cosh(c+d) - 2e^{-b}\cos(c-d)\right)$$

$$v_{[C_{4,2}L^1H^3{}_{(j-1,k-1,j-1)}]}C(b,c,d) = \frac{1}{4}\left(2e^b \sinh(c+d) - 2e^{-b}\sin(c-d)\right)$$

$$v_{[C_{4,2}L^1H^3{}_{(j-1,k-1,j-1)}]}D(b,c,d) = \frac{1}{4}\left(2e^b \sinh(c+d) + 2e^{-b}\sin(c-d)\right)$$

(10.23)

which are the same (variables have swapped about and names have swapped about) as the v-functions of the $C_{4,1}L^1H^3{}_{(j=1,k=1,l=1)}$ algebra.

$$vA - vB = e^{-b}\cos(c-d)$$
$$vC - vD = -e^{-b}\sin(c-d)$$
(10.24)

The v-functions - $C_{4,1}L^1E^3{}_{(j=1,k=1,l=1)}$:

The v-functions of the $C_{4,1}L^1E^3{}_{(j=-1,k=1,l=1)}$ algebra are:

$$v_{[C_{4,1}L^1E^3{}_{(j-1,k-1,j-1)}]}A(b,c,d) = \frac{1}{2}\left(e^{\left(\frac{b-d}{\sqrt{2}}\right)}\cos\left(\frac{b+d}{\sqrt{2}}+c\right) + e^{\left(\frac{b-d}{\sqrt{2}}\right)}\cos\left(\frac{b+d}{\sqrt{2}}-c\right)\right)$$

$$v_{[C_{4,1}L^1E^3{}_{(j-1,k-1,j-1)}]}B(b,c,d) = \frac{\sqrt{2}}{4}\left(\begin{array}{l}e^{\left(\frac{b-d}{\sqrt{2}}\right)}\left(\cos\left(\frac{b+d}{\sqrt{2}}+c\right)+\sin\left(\frac{b+d}{\sqrt{2}}+c\right)\right)\\+e^{\left(\frac{b-d}{\sqrt{2}}\right)}\left(-\cos\left(\frac{b+d}{\sqrt{2}}-c\right)+\sin\left(\frac{b+d}{\sqrt{2}}-c\right)\right)\end{array}\right)$$

(10.25)

$$v_{\left[C_a,LE'_{(j-1,k-1,j-1)}\right]}C(b,c,d) = \frac{1}{2}\left(e^{\left(\frac{b-d}{\sqrt{2}}\right)}\sin\left(\frac{b+d}{\sqrt{2}}+c\right) - e^{\left(\frac{b-d}{\sqrt{2}}\right)}\sin\left(\frac{b+d}{\sqrt{2}}-c\right)\right)$$

$$v_{\left[C_a,LE'_{(j-1,k-1,j-1)}\right]}D(b,c,d) = \frac{\sqrt{2}}{4}\left(\begin{array}{c}e^{\left(\frac{b-d}{\sqrt{2}}\right)}\left(-\cos\left(\frac{b+d}{\sqrt{2}}+c\right)+\sin\left(\frac{b+d}{\sqrt{2}}+c\right)\right)\\ +e^{\left(\frac{b-d}{\sqrt{2}}\right)}\left(\cos\left(\frac{b+d}{\sqrt{2}}-c\right)+\sin\left(\frac{b+d}{\sqrt{2}}-c\right)\right)\end{array}\right)$$

(10.26)

We have:

$$v_{\left[C_a,LE'_{(j-1,k-1,j-1)}\right]}A(0,0,d) = \cos\left(\frac{d}{\sqrt{2}}\right)\cosh\left(\frac{d}{\sqrt{2}}\right)$$

$$v_{\left[C_a,LE'_{(j-1,k-1,j-1)}\right]}B(0,0,d) = \frac{1}{\sqrt{2}}\left(\sin\left(\frac{d}{\sqrt{2}}\right)\cosh\left(\frac{d}{\sqrt{2}}\right) - \cos\left(\frac{d}{\sqrt{2}}\right)\sinh\left(\frac{d}{\sqrt{2}}\right)\right)$$

$$v_{\left[C_a,LE'_{(j-1,k-1,j-1)}\right]}C(0,0,d) = -\sinh\left(\frac{d}{\sqrt{2}}\right)\sin\left(\frac{d}{\sqrt{2}}\right)$$

$$v_{\left[C_a,LE'_{(j-1,k-1,j-1)}\right]}D(0,0,d) = \frac{1}{\sqrt{2}}\left(\cos\left(\frac{d}{\sqrt{2}}\right)\sinh\left(\frac{d}{\sqrt{2}}\right) + \sin\left(\frac{d}{\sqrt{2}}\right)\cosh\left(\frac{d}{\sqrt{2}}\right)\right)$$

(10.27)

And:

$$\begin{aligned}v_{\left[C_{4,1}LE^3_{(j-1,k-1,j-1)}\right]}A(0,c,0) &= \cos(c)\\ v_{\left[C_{4,1}LE^3_{(j-1,k-1,j-1)}\right]}B(0,c,0) &= 0\\ v_{\left[C_{4,1}LE^3_{(j-1,k-1,j-1)}\right]}C(0,c,0) &= \sin(c)\\ v_{\left[C_{4,1}LE^3_{(j-1,k-1,j-1)}\right]}D(0,c,0) &= 0\end{aligned}$$

(10.28)

And:

$$v_{\left[C_a,LE'_{(j-1,k-1,j-1)}\right]}A(b,0,0) = \cosh\left(\frac{b}{\sqrt{2}}\right)\cos\left(\frac{b}{\sqrt{2}}\right)$$

$$v_{\left[C_a,LE'_{(j-1,k-1,j-1)}\right]}B(b,0,0) = \frac{1}{\sqrt{2}}\left(\cos\left(\frac{b}{\sqrt{2}}\right)\sinh\left(\frac{b}{\sqrt{2}}\right) + \sin\left(\frac{b}{\sqrt{2}}\right)\cosh\left(\frac{b}{\sqrt{2}}\right)\right)$$

$$v_{\left[C_a,LE'_{(j-1,k-1,j-1)}\right]}C(b,0,0) = \sinh\left(\frac{b}{\sqrt{2}}\right)\sin\left(\frac{b}{\sqrt{2}}\right)$$

$$v_{\left[C_a,LE'_{(j-1,k-1,j-1)}\right]}D(b,0,0) = \frac{1}{\sqrt{2}}\left(-\cos\left(\frac{b}{\sqrt{2}}\right)\sinh\left(\frac{b}{\sqrt{2}}\right) + \sin\left(\frac{b}{\sqrt{2}}\right)\cosh\left(\frac{b}{\sqrt{2}}\right)\right)$$

(10.29)

Differentiation:

Section 10

$$\frac{\partial \left(v_{[\ 1]}A(b,c,d)\right)}{\partial b} = -v_{[\ 1]}D(b,c,d)$$

$$\frac{\partial \left(v_{[\ 1]}A(b,c,d)\right)}{\partial c} = -v_{[\ 1]}C(b,c,d) \tag{10.30}$$

$$\frac{\partial \left(v_{[\ 1]}A(b,c,d)\right)}{\partial d} = -v_{[\ 1]}B(b,c,d)$$

$$\frac{\partial \left(v_{[\ 1]}B(b,c,d)\right)}{\partial b} = v_{[\ 1]}A(b,c,d)$$

$$\frac{\partial \left(v_{[\ 1]}B(b,c,d)\right)}{\partial c} = -v_{[\ 1]}D(b,c,d) \tag{10.31}$$

$$\frac{\partial \left(v_{[\ 1]}B(b,c,d)\right)}{\partial d} = -v_{[\ 1]}C(b,c,d)$$

$$\frac{\partial \left(v_{[\ 1]}C(b,c,d)\right)}{\partial b} = v_{[\ 1]}B(b,c,d)$$

$$\frac{\partial \left(v_{[\ 1]}C(b,c,d)\right)}{\partial c} = v_{[\ 1]}A(b,c,d) \tag{10.32}$$

$$\frac{\partial \left(v_{[\ 1]}C(b,c,d)\right)}{\partial d} = -v_{[\ 1]}D(b,c,d)$$

$$\frac{\partial \left(v_{[\ 1]}D(b,c,d)\right)}{\partial b} = v_{[\ 1]}C(b,c,d)$$

$$\frac{\partial \left(v_{[\ 1]}D(b,c,d)\right)}{\partial c} = v_{[\ 1]}B(b,c,d) \tag{10.33}$$

$$\frac{\partial \left(v_{[\ 1]}D(b,c,d)\right)}{\partial d} = v_{[\ 1]}A(b,c,d)$$

As with the v-functions of the other algebras we have met, the v-functions of the $C_{4,1}L'E^3_{(j=-1,k=1,l=1)}$ algebra also have series expansions which have v-functions for the coefficients.

Section 11

2nd edition note: With hindsight, it might have been better to put this chapter at the very start of the book. Modern practice when working in this area of mathematics is to start with the finite group written as a set of permutation matrices and then form the algebra.

Introduction to Section 11:

In this section, we explore the connection between the natural algebras and the abelian groups. We are already aware that the algebraic matrix forms are copies of the standard form Cayley tables of the abelian groups. In this section, we find out that, in a profound sense, they are the abelian groups. What we have here is a unification of abelian groups and the natural algebraic fields. Since the natural algebraic fields contain the natural spaces, this is a unification of abelian groups, natural algebraic fields and natural geometries.

The unification is two-fold more extensive than this. Firstly, the natural algebraic fields are the vector product algebras. Secondly, we have abandoned the concept of algebraic extension of fields in favour of algebraic inclusion. This is effectively declaring that each of the natural algebraic fields is a different type of number. Thus we have it that each type of number is also an abelian group and also a natural algebra and also a vector product algebra and also a natural space. This is the unification of numbers, groups, algebras, vectors, and geometries – all the same one thing.

Permutation Matrices:

Square matrices whose elements are either 1 or 0 and in which these elements are arranged so that a 1 occurs once and only once in each row and in each column are called permutation matrices. We give some examples of 4×4 permutation matrices.

$$A : \begin{pmatrix} 1234 \\ 2341 \end{pmatrix} = (1234) \mapsto \begin{bmatrix} 0 & 1 & 0 & 0 \\ 0 & 0 & 1 & 0 \\ 0 & 0 & 0 & 1 \\ 1 & 0 & 0 & 0 \end{bmatrix} \qquad (11.1)$$

Section 11

$$B: \begin{pmatrix} 1234 \\ 3421 \end{pmatrix} = (1324) \mapsto \begin{bmatrix} 0 & 0 & 1 & 0 \\ 0 & 0 & 0 & 1 \\ 0 & 1 & 0 & 0 \\ 1 & 0 & 0 & 0 \end{bmatrix} \quad (11.2)$$

$$C: \begin{pmatrix} 1234 \\ 2413 \end{pmatrix} = (1243) \mapsto \begin{bmatrix} 0 & 1 & 0 & 0 \\ 0 & 0 & 0 & 1 \\ 1 & 0 & 0 & 0 \\ 0 & 0 & 1 & 0 \end{bmatrix} \quad (11.3)$$

$$D: \begin{pmatrix} 1234 \\ 4132 \end{pmatrix} = (142)(3) \mapsto \begin{bmatrix} 0 & 0 & 0 & 1 \\ 1 & 0 & 0 & 0 \\ 0 & 0 & 1 & 0 \\ 0 & 1 & 0 & 0 \end{bmatrix}$$

$$\quad (11.4)$$

$$E: \begin{pmatrix} 1234 \\ 2143 \end{pmatrix} = (12)(34) \mapsto \begin{bmatrix} 0 & 1 & 0 & 0 \\ 1 & 0 & 0 & 0 \\ 0 & 0 & 0 & 1 \\ 0 & 0 & 1 & 0 \end{bmatrix}$$

Convention has it that all of the above matrices are permutation matrices of equal status. We differ in this view and we refer to the unbroken permutations $\{A, B, C\}$ as pure permutations and the broken permutations as impure permutations.

Nomenclature:
Pure Permutation Matrices
A pure permutation matrix is a permutation matrix that is associated with an unbroken permutation cycle.

Permutation matrices are made by swapping columns of the identity matrix. Since the identity matrix is non-singular and non-singularity is maintained under column swaps, permutation matrices are always non-singular. The inverse of a permutation matrix corresponds to the inverse permutation.

Since every finite group can be represented as a permutation group, it is of no great surprise that permutation matrices are associated with groups. The unbroken permutations above, $\{B, C, D\}$, are associated with the cyclic group, C_4.

The broken permutation, D, is associated with 'fixing together' the groups $\{C_1, C_3\}$.

$$\begin{bmatrix} 0 & 0 & 0 & 1 \\ 1 & 0 & 0 & 0 \\ 0 & 0 & 1 & 0 \\ 0 & 1 & 0 & 0 \end{bmatrix} \mapsto \begin{bmatrix} \begin{bmatrix} 0 & 1 & 0 \\ 0 & 0 & 1 \\ 1 & 0 & 0 \end{bmatrix} & \begin{matrix} 0 \\ 0 \\ 0 \end{matrix} \\ \begin{matrix} 0 & 0 & 0 \end{matrix} & [1] \end{bmatrix} \qquad (11.5)$$

The presence of a 1 on the leading diagonal betrays the presence of the C_1 group. Of course, 'fixing together' of groups is an impure thing. What we have here is the 3×3 matrix of a pure permutation and the 1×1 matrix of a pure permutation fitted on to the leading diagonal of a 4×4 matrix. If we were to form the direct product of the groups $\{C_1, C_3\}$, we would get the group C_3, and this is associated with 3×3 permutation matrices. Such 'fixed together' permutation matrices are not our main concern.

All the permutation matrices generate (by matrix multiplication – we raise them to various powers of themselves) the group (or 'fixing together' of groups) with which they are associated. We demonstrate with the 1×1 permutation matrices (every one of them), the 2×2 permutation matrices (every one of them), and the 3×3 permutation matrices (every one of them). The powers of these matrices are:

$$[1]^1 = [1] \qquad (11.6)$$

$$\begin{bmatrix} 1 & 0 \\ 0 & 1 \end{bmatrix}^1 = \begin{bmatrix} 1 & 0 \\ 0 & 1 \end{bmatrix}$$

$$\begin{bmatrix} 0 & 1 \\ 1 & 0 \end{bmatrix}^{1\ldots 2} = \begin{bmatrix} 0 & 1 \\ 1 & 0 \end{bmatrix}, \begin{bmatrix} 1 & 0 \\ 0 & 1 \end{bmatrix} \qquad (11.7)$$

$$F : \begin{bmatrix} 1 & 0 & 0 \\ 0 & 1 & 0 \\ 0 & 0 & 1 \end{bmatrix}^1 = \begin{bmatrix} 1 & 0 & 0 \\ 0 & 1 & 0 \\ 0 & 0 & 1 \end{bmatrix} : C_1 \qquad (11.8)$$

Section 11

$$G : \begin{bmatrix} 1 & 0 & 0 \\ 0 & 0 & 1 \\ 0 & 1 & 0 \end{bmatrix}^{1.2} = \begin{bmatrix} 1 & 0 & 0 \\ 0 & 0 & 1 \\ 0 & 1 & 0 \end{bmatrix}, \begin{bmatrix} 1 & 0 & 0 \\ 0 & 1 & 0 \\ 0 & 0 & 1 \end{bmatrix} : C_2 \times C_1$$

$$H : \begin{bmatrix} 0 & 1 & 0 \\ 1 & 0 & 0 \\ 0 & 0 & 1 \end{bmatrix}^{1.2} = \begin{bmatrix} 0 & 1 & 0 \\ 1 & 0 & 0 \\ 0 & 0 & 1 \end{bmatrix}, \begin{bmatrix} 1 & 0 & 0 \\ 0 & 1 & 0 \\ 0 & 0 & 1 \end{bmatrix} : C_2 \times C_1$$

(11.9)

And:

$$I : \begin{bmatrix} 0 & 1 & 0 \\ 0 & 0 & 1 \\ 1 & 0 & 0 \end{bmatrix}^{1.3} = \begin{bmatrix} 0 & 1 & 0 \\ 0 & 0 & 1 \\ 1 & 0 & 0 \end{bmatrix}, \begin{bmatrix} 0 & 0 & 1 \\ 1 & 0 & 0 \\ 0 & 1 & 0 \end{bmatrix}, \begin{bmatrix} 1 & 0 & 0 \\ 0 & 1 & 0 \\ 0 & 0 & 1 \end{bmatrix} : C_3$$

$$J : \begin{bmatrix} 0 & 0 & 1 \\ 1 & 0 & 0 \\ 0 & 1 & 0 \end{bmatrix}^{1.3} = \begin{bmatrix} 0 & 0 & 1 \\ 1 & 0 & 0 \\ 0 & 1 & 0 \end{bmatrix}, \begin{bmatrix} 0 & 1 & 0 \\ 0 & 0 & 1 \\ 1 & 0 & 0 \end{bmatrix}, \begin{bmatrix} 1 & 0 & 0 \\ 0 & 1 & 0 \\ 0 & 0 & 1 \end{bmatrix} : C_3 \quad (11.10)$$

$$K : \begin{bmatrix} 0 & 0 & 1 \\ 0 & 1 & 0 \\ 1 & 0 & 0 \end{bmatrix}^{1.2} = \begin{bmatrix} 0 & 0 & 1 \\ 0 & 1 & 0 \\ 1 & 0 & 0 \end{bmatrix}, \begin{bmatrix} 1 & 0 & 0 \\ 0 & 1 & 0 \\ 0 & 0 & 1 \end{bmatrix} : C_2$$

Of all the 3×3 permutation matrices, only $\{I, J\}$ are pure permutation matrices. The other matrices are 3×3 matrices with lesser matrices fitted inside them along the leading diagonal. Pure permutation matrices have a leading diagonal that is all zeros.

In the case of 3×3 permutation matrices, we select the three powers of either of the pure permutation matrices $\{I, J\}$ - they are the same matrices.

$$I^1 = \begin{bmatrix} 0 & 1 & 0 \\ 0 & 0 & 1 \\ 1 & 0 & 0 \end{bmatrix}, I^2 = \begin{bmatrix} 0 & 0 & 1 \\ 1 & 0 & 0 \\ 0 & 1 & 0 \end{bmatrix}, I^3 = \begin{bmatrix} 1 & 0 & 0 \\ 0 & 1 & 0 \\ 0 & 0 & 1 \end{bmatrix} \quad (11.11)$$

We change them into variable matrices:

$$B = \begin{bmatrix} 0 & b & 0 \\ 0 & 0 & b \\ b & 0 & 0 \end{bmatrix}, C = \begin{bmatrix} 0 & 0 & c \\ c & 0 & 0 \\ 0 & c & 0 \end{bmatrix}, A = \begin{bmatrix} a & 0 & 0 \\ 0 & a & 0 \\ 0 & 0 & a \end{bmatrix} \quad (11.12)$$

We add them:

$$\begin{bmatrix} a & b & c \\ c & a & b \\ b & c & a \end{bmatrix} = C_3 L^1 H^2{}_{(j=1,k=1)} \quad (11.13)$$

Thus, from the set of 3×3 permutation matrices, we have constructed the basic layout of the 3-dimensional algebraic matrix form[106].

In the case of the 4×4 permutation matrices, there are nine permutation matrices that have only zeros on the leading diagonal. They come in sets of threes (two C_4 s and one C_2 [107]) which, when turned into variables and added produce the three 4-dimensional algebraic matrix forms $\{C_{4.1} L^1 H^3, C_{4.2} L^1 H^3, C_{4.3} L^1 H^3\}$, and so these algebras are associated with the group C_4.

There is one other way of combining three of these nine (all zeros on the leading diagonal) 4×4 permutation matrices; the three C_2 permutation matrices can be combined together, and this gives the $C_2 \times C_2 L^1 H^3$ algebraic matrix form, and so this algebra is thus associated with the only other abelian group of order 4, $C_2 \times C_2$.

The group C_4 has three different, but algebraically isomorphic, 4×4 algebraic matrix forms associated with it. In one of these, $C_{4.1} L^1 H^3$, the C_2 sub-group is generated by the c element; in the $C_{4.2} L^1 H^3$ algebra, the C_2 sub-group is generated by the b element; in the $C_{4.3} L^1 H^3$ algebra, the C_2 sub-group is generated by the d element.

---

[106] And the Cayley table of the order 3 cyclic group, of course.

[107] C_4 has a C_2 sub-group.

Section 11

Perusal of the nine 4×4 permutation matrices that have only zeros on the leading diagonal will give the reader a clearer understanding of the basis of algebraic isomorphism and its origins in these algebras.

We have now established a clear connection between the three and four dimensional algebraic matrix forms and the finite abelian groups of order three and four. The connection expresses itself very succinctly in the Cayley tables of the finite abelian groups. For example, the standard form Cayley table of the group C_3 is:

$$\begin{bmatrix} A & B & C \\ C & A & B \\ B & C & A \end{bmatrix} \qquad (11.14)$$

which, if copied into a matrix, forms the $C_3 L^1 H^2$ algebraic matrix form:

$$C_3 L^1 H^2 = \begin{bmatrix} a & b & c \\ c & a & b \\ b & c & a \end{bmatrix} \qquad (11.15)$$

We expect that this phenomenon is general, and we can write down the algebraic matrix form associated with any finite abelian group by copying the standard form Cayley table of that group into a matrix, but we can offer no proof that it is general. In short, we are elevating every finite abelian group to the status of algebraic field – well, perhaps not that exactly.[108]

Roots of Unity:
The eigenvalues of a permutation matrix are the roots of unity associated with the size of the sub-matrices. For example consider the permutation matrix:

---

[108] The same works with non-abelian finite groups, but one gets a multiplicatively non-commutative algebra. These algebras usually require to be restricted (the quaternions are exceptional), and it might be impossible to find a polar form of such algebras – your author conjectures that no polar forms exist for these algebras. Your author has often been wrong.

$$P_\sigma = \begin{bmatrix} 0 & 0 & 0 & 0 & 1 & 0 \\ 1 & 0 & 0 & 0 & 0 & 0 \\ 0 & 0 & 0 & 0 & 0 & 1 \\ 0 & 0 & 0 & 1 & 0 & 0 \\ 0 & 1 & 0 & 0 & 0 & 0 \\ 0 & 0 & 1 & 0 & 0 & 0 \end{bmatrix} \equiv \begin{bmatrix} \begin{bmatrix} 0 & 1 & 0 \\ 0 & 0 & 1 \\ 1 & 0 & 0 \end{bmatrix} & 0 & 0 & 0 \\ 0 & 0 & 0 \\ 0 & 0 & 0 \\ 0 & 0 & 0 & [1] & 0 & 0 \\ 0 & 0 & 0 & 0 & \begin{bmatrix} 0 & 1 \\ 1 & 0 \end{bmatrix} \end{bmatrix}$$ (11.16)

The characteristic polynomial of this matrix is: $(\lambda^3 - 1)(\lambda - 1)(\lambda^2 - 1) = 0$, and so it has eigenvalues $\{1^{\frac{1}{3}}, 1^{\frac{1}{3}}, 1^{\frac{1}{3}}, 1^1, 1^{\frac{1}{2}}, 1^{\frac{1}{2}}\}$. Similarly, the characteristic polynomial of the $\{A, B, C\}$ matrices above is: $(\lambda^4 - 1) = 0$, and they have eigenvalues $\{1^{\frac{1}{4}}, 1^{\frac{1}{4}}, 1^{\frac{1}{4}}, 1^{\frac{1}{4}}\}$. Since the set of powers of the pure permutation matrices are used to make the algebraic matrix forms, the $n \times n$ algebraic matrix forms will be associated with the nth roots of unity. We have seen this phenomenon in the non-matrix notation of the algebraic matrix forms.

However, these roots of unity will not be the Euclidean complex number roots of unity. They are the roots of unity in algebras different from the Euclidean complex numbers. In the case of the C_3 algebras, they are $\{1, \hat{p}, \hat{q}\}$. In the case of the C_4 algebras, they are $\{1, \hat{r}, \hat{s}, \hat{t}\}$. Even so, to every root of unity in an algebra other than the Euclidean complex numbers, there corresponds a root of unity in the Euclidean complex numbers, and so it is that the other algebras can be shadowed by the Euclidean complex numbers.

Algebraic Isomorphism between Algebraic Matrix Forms:

We have previously seen that the three algebraically isomorphic forms of the C_4 algebras differ from each other in that the element forming the C_2 subgroup is respectively $\{C_{4.1} : c, C_{4.2} : b, C_{4.3} : d\}$. The reader might think that only groups with sub-groups have more than one (isomorphic) algebraic matrix form. This is not the case. The number of isomorphic forms depends upon the number of pure permutation matrices of a particular size, as we demonstrate below.

Section 11

The 5-dimensional algebraic matrix form that is a copy of the standard form of the Cayley Table of C_5 is:

$$C_{5.1}L^1H^4{}_{(j=1,k=1,l=1,m=1)} = \begin{bmatrix} a & b & c & d & e \\ e & a & b & c & d \\ d & e & a & b & c \\ c & d & e & a & b \\ b & c & d & e & a \end{bmatrix} \qquad (11.17)$$

If we raise a particular sub-matrix through integer powers, we get the other unit sub-matrices in a particular order. For example, the b sub-matrix of this algebra is:

$$B = \begin{bmatrix} 0 & b & 0 & 0 & 0 \\ 0 & 0 & b & 0 & 0 \\ 0 & 0 & 0 & b & 0 \\ 0 & 0 & 0 & 0 & b \\ b & 0 & 0 & 0 & 0 \end{bmatrix} \qquad (11.18)$$

Raised to integer powers, this matrix becomes:

$$B^{1\ldots5} = B, C, D, E, A \qquad (11.19)$$

There are other 5-dimensioal algebraic matrix forms. They are:

$$\left\{ C_{5.2}L^1H^4 = \begin{bmatrix} a & b & c & d & e \\ c & a & d & e & b \\ b & e & a & c & d \\ e & d & b & a & c \\ d & c & e & b & a \end{bmatrix}, C_{5.3}L^1H^4 = \begin{bmatrix} a & b & c & d & e \\ c & a & e & b & d \\ b & d & a & e & c \\ e & c & d & a & b \\ d & e & b & c & a \end{bmatrix} \right\}$$

$$(11.20)$$

And:

174

$$C_{5,4}L^{1}H^{4} = \begin{bmatrix} a & b & c & d & e \\ d & a & e & c & b \\ e & c & a & b & d \\ b & e & d & a & c \\ c & d & b & e & a \end{bmatrix}$$

$$C_{5,5}L^{1}H^{4} = \begin{bmatrix} a & b & c & d & e \\ e & a & d & b & c \\ d & c & a & e & b \\ c & e & b & a & d \\ b & d & e & c & a \end{bmatrix}, C_{5,6}L^{1}H^{4} = \begin{bmatrix} a & b & c & d & e \\ d & a & b & e & c \\ e & d & a & c & b \\ b & c & e & a & d \\ c & e & d & b & a \end{bmatrix}$$

(11.21)

The respective b sub-matrix structures are:

$$L^{1}H^{4}_{2} : B^{1...5} = B, E, D, C, A$$
$$L^{1}H^{4}_{3} : B^{1...5} = B, D, E, C, A$$
$$L^{1}H^{4}_{4} : B^{1...5} = B, E, C, D, A \qquad (11.22)$$
$$L^{1}H^{4}_{5} : B^{1...5} = B, D, C, E, A$$
$$L^{1}H^{4}_{6} : B^{1...5} = B, C, E, D, A$$

So, even though the cyclic group of order five has no proper sub-groups, it does have six algebraically isomorphic algebraic matrix forms. There are $4! = 24$ 5×5 permutation matrices that have all zeros on the leading diagonal and which generate the group C_5.[109] These are taken together in sets of four to make an algebraic matrix form. Hence six algebraically isomorphic algebraic matrix forms.

Thus, our algebraic matrix forms are only a little more than abelian groups.[110] This is the reason why we named the algebras after the groups that underlie them.

---

[109] There are other 5×5 permutation matrices with zeros on the leading diagonal. These matrices are effectively direct products of 2×2 and 3×3 permutation matrices.

[110] We use the word 'little' in the large sense.

Section 11

Permutation Matrices with Negative Entries:

Although convention defines permutation matrices to have entries that are all unity, they can be written in smaller (half-size) matrices with entries that are unity or minus unity. The matrix:

$$B^1 = \begin{bmatrix} 0 & 0 & 1 & 0 \\ 0 & 0 & 0 & 1 \\ 0 & 1 & 0 & 0 \\ 1 & 0 & 0 & 0 \end{bmatrix} \quad (11.23)$$

is a permutation matrix that generates three other permutation matrices when raised to different powers. (It is a pure permutation matrix.)

$$B^2 = \begin{bmatrix} 0 & 1 & 0 & 0 \\ 1 & 0 & 0 & 0 \\ 0 & 0 & 0 & 1 \\ 0 & 0 & 1 & 0 \end{bmatrix}, B^3 = \begin{bmatrix} 0 & 0 & 0 & 1 \\ 0 & 0 & 1 & 0 \\ 1 & 0 & 0 & 0 \\ 0 & 1 & 0 & 0 \end{bmatrix}, B^4 = \begin{bmatrix} 1 & 0 & 0 & 0 \\ 0 & 1 & 0 & 0 \\ 0 & 0 & 1 & 0 \\ 0 & 0 & 0 & 1 \end{bmatrix}$$
(11.24)

The matrix:

$$C^1 = \begin{bmatrix} 0 & 1 \\ -1 & 0 \end{bmatrix} \quad (11.25)$$

also generates three matrices:

$$C^2 = \begin{bmatrix} -1 & 0 \\ 0 & -1 \end{bmatrix}, C^3 = \begin{bmatrix} 0 & -1 \\ 1 & 0 \end{bmatrix}, C^4 = \begin{bmatrix} 1 & 0 \\ 0 & 1 \end{bmatrix} \quad (11.26)$$

In both cases, the set of matrices forms the group C_4 under the operation of matrix multiplication. There is a one-to-one and onto mapping between these two sets of matrices.

$$B^1: \begin{bmatrix} 0 & 0 & 1 & 0 \\ 0 & 0 & 0 & 1 \\ 0 & 1 & 0 & 0 \\ 1 & 0 & 0 & 0 \end{bmatrix} \to C^1: \begin{bmatrix} 0 & 1 \\ -1 & 0 \end{bmatrix}, \quad B^2: \begin{bmatrix} 0 & 1 & 0 & 0 \\ 1 & 0 & 0 & 0 \\ 0 & 0 & 0 & 1 \\ 0 & 0 & 1 & 0 \end{bmatrix} \to C^2: \begin{bmatrix} -1 & 0 \\ 0 & -1 \end{bmatrix}$$

$$B^3: \begin{bmatrix} 0 & 0 & 0 & 1 \\ 0 & 0 & 1 & 0 \\ 1 & 0 & 0 & 0 \\ 0 & 1 & 0 & 0 \end{bmatrix} \to C^3: \begin{bmatrix} 0 & -1 \\ 1 & 0 \end{bmatrix}, \quad B^4: \begin{bmatrix} 1 & 0 & 0 & 0 \\ 0 & 1 & 0 & 0 \\ 0 & 0 & 1 & 0 \\ 0 & 0 & 0 & 1 \end{bmatrix} \to C^4: \begin{bmatrix} 1 & 0 \\ 0 & 1 \end{bmatrix}$$

(11.27)

These two sets of matrices are isomorphic over the operation of multiplication, as can be shown by multiplying together every pair within the set and comparing the result to the same product within the other set. Furthermore, by similar comparison of sums of pairs of matrices within a set, these two sets of matrices are isomorphic over the operation of addition. Equivalence under scalar multiplication is trivial.

Theorem:
The above two sets of matrices are algebraically equivalent.
Proof:
    See above

Since algebraic matrix forms are constructed by turning a suitable set of permutation matrices into variable matrices and adding them. We have:

$$\begin{bmatrix} 0 & 0 & c & 0 \\ 0 & 0 & 0 & c \\ 0 & c & 0 & 0 \\ c & 0 & 0 & 0 \end{bmatrix} + \begin{bmatrix} 0 & b & 0 & 0 \\ b & 0 & 0 & 0 \\ 0 & 0 & 0 & b \\ 0 & 0 & b & 0 \end{bmatrix}$$

$$+ \begin{bmatrix} 0 & 0 & 0 & d \\ 0 & 0 & d & 0 \\ d & 0 & 0 & 0 \\ 0 & d & 0 & 0 \end{bmatrix} + \begin{bmatrix} a & 0 & 0 & 0 \\ 0 & a & 0 & 0 \\ 0 & 0 & a & 0 \\ 0 & 0 & 0 & a \end{bmatrix} = C_{4,2} L^1 H^3_{(j-1,k-1,l-1)}$$

(11.28)

$$\begin{bmatrix} 0 & c \\ -c & 0 \end{bmatrix} + \begin{bmatrix} -b & 0 \\ 0 & -b \end{bmatrix} + \begin{bmatrix} 0 & -d \\ d & 0 \end{bmatrix} + \begin{bmatrix} a & 0 \\ 0 & a \end{bmatrix}$$

$$= \begin{bmatrix} a-b & c-d \\ -(c-d) & a-b \end{bmatrix} = \mathbb{C}(a-b, c-d)$$

(11.29)

177

Section 11

This is, of course, the algebraic equivalence of the Euclidean complex numbers and the $C_{4.2}L^1H^3_{(j=1,k=1,l=1)}$ algebra.

$$\begin{bmatrix} a & b & c & d \\ b & a & d & c \\ d & c & a & b \\ c & d & b & a \end{bmatrix} \cong \begin{bmatrix} a-b & c-d \\ -(c-d) & a-b \end{bmatrix} \quad (11.30)$$

Of course, algebraic isomorphism does not mean that two algebras are identical in all ways.

Complex Numbers The Higher Dimensional Forms – 2nd Edition

Section 12

Introduction to Section 12:

In this section, we introduce the reader to a new matrix operation, which we call folding and unfolding a matrix. This operation is not an algebraic operation like addition and multiplication. It is only a notation operation that changes from non-matrix notation to matrix notation. However, when applied to the negative elements of algebraic matrix forms with negative space squeezing parameters, it unfolds the algebra into a higher dimensional algebra. This is quite startling. It is also contentious. Whether or not the folding and unfolding operation is allowable is a matter for the reader and their conscience.

The Folding Operation:

It is a basic property of matrices that they can be partitioned into blocks and multiplied together as if the blocks were the elements of a matrix. When this is done, the result is the same as if the matrices were not partitioned. We give an example:

$$\begin{bmatrix} a & b & c & d \\ e & f & g & h \\ i & j & k & l \\ m & n & o & p \end{bmatrix} \begin{bmatrix} \alpha & \beta & \chi & \delta \\ \varepsilon & \phi & \varphi & \gamma \\ \eta & \iota & \kappa & \lambda \\ \mu & \nu & o & \pi \end{bmatrix} =$$

$$\begin{bmatrix} a\alpha + b\varepsilon + c\eta + d\mu & a\beta + b\phi + c\iota + d\nu & a\chi + b\varphi + c\kappa + do & a\delta + b\lambda + c\lambda + d\pi \\ e\alpha + f\varepsilon + g\eta + h\mu & e\beta + f\phi + g\iota + h\nu & e\chi + f\varphi + g\kappa + ho & e\delta + f\lambda + g\lambda + h\pi \\ i\alpha + j\varepsilon + k\eta + l\mu & i\beta + j\phi + k\iota + l\nu & i\chi + j\varphi + k\kappa + lo & i\delta + j\lambda + k\lambda + l\pi \\ m\alpha + n\varepsilon + o\eta + p\mu & m\beta + n\phi + o\iota + p\nu & m\chi + n\varphi + o\kappa + po & m\delta + n\lambda + o\lambda + p\pi \end{bmatrix}$$

(12.1)

2nd edition note: The unfolding operation has small but useful application in opening a matrix with complex elements into a matrix with real elements whereby we can see what is really happening. We do not need hermitian matrices in quantum mechanics and can unfold them. We can similarly unfold the standard $SU(2)$ matrix and see that it is a quaternion.

Section 12

$$\begin{bmatrix} \begin{bmatrix} a & b \\ e & f \end{bmatrix} & \begin{bmatrix} c & d \\ g & h \end{bmatrix} \\ \begin{bmatrix} i & j \\ m & n \end{bmatrix} & \begin{bmatrix} k & l \\ o & p \end{bmatrix} \end{bmatrix} \begin{bmatrix} \begin{bmatrix} \alpha & \beta \\ \varepsilon & \phi \end{bmatrix} & \begin{bmatrix} \chi & \delta \\ \varphi & \lambda \end{bmatrix} \\ \begin{bmatrix} \eta & \iota \\ \mu & \nu \end{bmatrix} & \begin{bmatrix} \kappa & \lambda \\ o & \pi \end{bmatrix} \end{bmatrix} =$$

$$\begin{bmatrix} \begin{bmatrix} a & b \\ e & f \end{bmatrix}\begin{bmatrix} \alpha & \beta \\ \varepsilon & \phi \end{bmatrix} + \begin{bmatrix} c & d \\ g & h \end{bmatrix}\begin{bmatrix} \eta & \iota \\ \mu & \nu \end{bmatrix} & \begin{bmatrix} a & b \\ e & f \end{bmatrix}\begin{bmatrix} \chi & \delta \\ \varphi & \lambda \end{bmatrix} + \begin{bmatrix} c & d \\ g & h \end{bmatrix}\begin{bmatrix} \kappa & \lambda \\ o & \pi \end{bmatrix} \\ \begin{bmatrix} i & j \\ m & n \end{bmatrix}\begin{bmatrix} \alpha & \beta \\ \varepsilon & \phi \end{bmatrix} + \begin{bmatrix} k & l \\ o & p \end{bmatrix}\begin{bmatrix} \eta & \iota \\ \mu & \nu \end{bmatrix} & \begin{bmatrix} i & j \\ m & n \end{bmatrix}\begin{bmatrix} \chi & \delta \\ \varphi & \lambda \end{bmatrix} + \begin{bmatrix} k & l \\ o & p \end{bmatrix}\begin{bmatrix} \kappa & \lambda \\ o & \pi \end{bmatrix} \end{bmatrix} =$$

$$\begin{bmatrix} \begin{bmatrix} a\alpha + b\varepsilon + c\eta + d\mu & a\beta + b\phi + c\iota + d\nu \\ e\alpha + f\varepsilon + g\eta + h\mu & e\beta + f\phi + g\iota + h\nu \end{bmatrix} & \begin{bmatrix} a\chi + b\varphi + c\kappa + do & a\delta + b\lambda + c\lambda + d\pi \\ e\chi + f\varphi + g\kappa + ho & e\delta + f\lambda + g\lambda + h\pi \end{bmatrix} \\ \begin{bmatrix} i\alpha + j\varepsilon + k\eta + l\mu & i\beta + j\phi + k\iota + l\nu \\ m\alpha + n\varepsilon + o\eta + p\mu & m\beta + n\phi + o\iota + p\nu \end{bmatrix} & \begin{bmatrix} i\chi + j\varphi + k\kappa + lo & i\delta + j\lambda + k\lambda + l\pi \\ m\chi + n\varphi + o\kappa + po & m\delta + n\lambda + o\lambda + p\pi \end{bmatrix} \end{bmatrix}$$

(12.2)

Block addition obviously gives the same results as normal addition. Thus, we have covered both the algebraic operations.

The matrix representation of Euclidean complex numbers is:

$$a + \hat{i}b \equiv \begin{bmatrix} a & b \\ -b & a \end{bmatrix} \quad (12.3)$$

Consider a matrix with complex numbers as elements; we can replace each complex number with a 2×2 matrix of the complex number form.

$$\begin{bmatrix} a+\hat{i}b & c+\hat{i}d \\ e+\hat{i}f & g-\hat{i}h \end{bmatrix} \triangleleft \begin{bmatrix} \begin{bmatrix} a & b \\ -b & a \end{bmatrix} & \begin{bmatrix} c & d \\ -d & c \end{bmatrix} \\ \begin{bmatrix} e & f \\ -f & e \end{bmatrix} & \begin{bmatrix} g & -h \\ h & g \end{bmatrix} \end{bmatrix} = \begin{bmatrix} a & b & c & d \\ -b & a & -d & c \\ e & f & g & -h \\ -f & e & h & g \end{bmatrix}$$

(12.4)

We refer to such expansion from a smaller matrix to a larger matrix as unfolding the matrix. We refer to the inverse operation as folding the matrix.

Notation: $\{\triangleleft, \triangleright\}$

Reading left to right, we use the $\triangleleft$ sign to indicate unfolding a smaller matrix into a larger matrix. Reading left to right, we use the $\triangleright$ sign to indicate folding a larger matrix into a smaller matrix.

The algebraic equivalence of normal matrices and blocked matrices guarantees that the large matrix is algebraically the same as the small matrix with the complex elements. We have effectively 'unfolded' the complex numbers into real numbers. The procedure is reversible if the 4×4 matrices contain blocks of the form of a complex number.

Folding and unfolding an algebraic matrix form in this way is no more than changing the notation. We have been doing this with complex numbers from the start of this work:

$$\begin{bmatrix} a + \hat{i}b \end{bmatrix} \triangleleft \begin{bmatrix} a & b \\ -b & a \end{bmatrix} \qquad (12.5)$$

We can unfold a single element into an overblown representation using the algebraic isomorphism of the real number matrices.

$$\begin{bmatrix} a \end{bmatrix} \triangleleft \begin{bmatrix} a & 0 \\ 0 & a \end{bmatrix} \qquad (12.6)$$

By this means, we obtain an overblown representation of an algebra.

$$\begin{bmatrix} a + \hat{i}b \end{bmatrix} \triangleleft \begin{bmatrix} a & b \\ -b & a \end{bmatrix} \triangleleft \begin{bmatrix} a & 0 & b & 0 \\ 0 & a & 0 & b \\ -b & 0 & a & 0 \\ 0 & -b & 0 & a \end{bmatrix} \qquad (12.7)$$

Again, algebraically, this is a notational change, but not all things (square roots) are algebraic operations.

We are not restricted to Euclidean complex numbers. Because the 4-dimensional C_{42} algebraic matrix form is four (blocks of) 2-dimensional numbers, it can be folded:

$$\begin{Bmatrix} C_{42} \overset{1}{L} E^3_{(j\ -1,k\ -1,l\ -1)} : \begin{bmatrix} a & b & c & d \\ -b & a & -d & c \\ d & -c & a & b \\ c & d & -b & a \end{bmatrix} \triangleright \begin{bmatrix} a + \hat{i}b & c + \hat{i}d \\ d - \hat{i}c & a + \hat{i}b \end{bmatrix} \triangleright a + s\hat{b} + t\hat{c} + u\hat{d} \\ \\ C_{42} \overset{1}{L} E^3_{(j\ -1,k\ -1,l\ -1)} : \begin{bmatrix} a & b & c & d \\ -b & a & -d & c \\ -d & c & a & b \\ -c & -d & -b & a \end{bmatrix} \triangleright \begin{bmatrix} a + \hat{i}b & c + \hat{i}d \\ -d + \hat{i}c & a + \hat{i}b \end{bmatrix} \triangleright a + s\hat{b} + t\hat{c} + u\hat{d} \end{Bmatrix}$$

(12.8)

Section 12

$$\left\{\begin{array}{l} C_{4.2}L^1H^3{}_{(j-1,k-1,l-1)}: \begin{bmatrix} a & b & c & d \\ b & a & d & c \\ -d & -c & a & b \\ -c & -d & b & a \end{bmatrix} \triangleright \begin{bmatrix} a+\hat{r}b & c+\hat{r}d \\ -d-\hat{r}c & a+\hat{r}b \end{bmatrix} \triangleright a+\hat{s}b+\hat{t}c+\hat{u}d \\ \\ C_{4.2}L^1H^3{}_{(j-1,k-1,l-1)}: \begin{bmatrix} a & b & c & d \\ b & a & d & c \\ d & c & a & b \\ c & d & b & a \end{bmatrix} \triangleright \begin{bmatrix} a+\hat{r}b & c+\hat{r}d \\ d+\hat{r}c & a+\hat{r}b \end{bmatrix} \triangleright a+\hat{s}b+\hat{t}c+\hat{u}d \end{array}\right.$$

(12.9)

where $\hat{r} = \sqrt{+1}$.

Calculation Example:

The $C_{4.2}L^1E^3{}_{(j=-1,k=-1,l=1)}$ algebra is:

$$\begin{bmatrix} a & b & c & d \\ -b & a & -d & c \\ d & -c & a & b \\ c & d & -b & a \end{bmatrix} \quad (12.10)$$

We have:

$$\left\{ \begin{bmatrix} a & 0 & 0 & 0 \\ 0 & a & 0 & 0 \\ 0 & 0 & a & 0 \\ 0 & 0 & 0 & a \end{bmatrix} \triangleright \begin{bmatrix} a & 0 \\ 0 & a \end{bmatrix}, \begin{bmatrix} 0 & b & 0 & 0 \\ -b & 0 & 0 & 0 \\ 0 & 0 & 0 & b \\ 0 & 0 & -b & 0 \end{bmatrix} \triangleright \begin{bmatrix} \hat{ib} & 0 \\ 0 & \hat{ib} \end{bmatrix}, \right.$$

$$\left. \begin{bmatrix} 0 & 0 & c & 0 \\ 0 & 0 & 0 & c \\ 0 & -c & 0 & 0 \\ c & 0 & 0 & 0 \end{bmatrix} \triangleright \begin{bmatrix} 0 & c \\ -\hat{ic} & 0 \end{bmatrix}, \begin{bmatrix} 0 & 0 & 0 & d \\ 0 & 0 & -d & 0 \\ d & 0 & 0 & 0 \\ 0 & d & 0 & 0 \end{bmatrix} \triangleright \begin{bmatrix} 0 & \hat{id} \\ d & 0 \end{bmatrix} \right\}$$

(12.11)

Adding the four 2×2 matrices together gives the 2×2 form of the $C_{4.2}L^1E^3{}_{(j=-1,k=-1,l=1)}$ algebra:

$$\begin{bmatrix} a+\hat{i}b & c+\hat{i}d \\ d-\hat{i}c & a+\hat{i}b \end{bmatrix} \qquad (12.12)$$

In non-matrix notation, the $C_{4,2}L^lE^3_{(j=-1,k=-1,l=1)}$ algebra is: $a+\hat{s}b+\hat{t}c+\hat{u}d$. Thus:

$$a+\hat{s}b+\hat{t}c+\hat{u}d \triangleleft \begin{bmatrix} a+\hat{i}b & c+\hat{i}d \\ d-\hat{i}c & a+\hat{i}b \end{bmatrix} \qquad (12.13)$$

Unfolding Negative Real Numbers:

The negative real numbers are not an algebra because they are not multiplicatively closed and they do not include the multiplicative identity. Matrices of the form:

$$\begin{bmatrix} 0 & b \\ b & 0 \end{bmatrix} \qquad (12.14)$$

have exactly the same algebraic failings. Furthermore, over the algebraic operations, these matrices are isomorphic to the negative real numbers.

Theorem:

Matrices of the form $\begin{bmatrix} 0 & b \\ b & 0 \end{bmatrix}$ are isomorphic to the negative real numbers over the algebraic operations.

Proof:

$$\phi:(-\mathbb{R},+,\times) \mapsto (M_{(2,\mathbb{R})},+,\bullet) \quad : \quad \phi([-b]) \mapsto \begin{bmatrix} 0 & b \\ b & 0 \end{bmatrix}$$

$$\phi([-c]+[-d]) \mapsto \begin{bmatrix} 0 & c \\ c & 0 \end{bmatrix} + \begin{bmatrix} 0 & d \\ d & 0 \end{bmatrix} = \begin{bmatrix} 0 & c+d \\ c+d & 0 \end{bmatrix}$$

$$\phi([-c]\times[-d]) \mapsto \begin{bmatrix} 0 & c \\ c & 0 \end{bmatrix} \bullet \begin{bmatrix} 0 & d \\ d & 0 \end{bmatrix} = \begin{bmatrix} cd & 0 \\ 0 & cd \end{bmatrix}$$

$$\phi(\alpha[-c]) \mapsto \begin{bmatrix} \alpha & 0 \\ 0 & \alpha \end{bmatrix} \begin{bmatrix} 0 & c \\ c & 0 \end{bmatrix} = \begin{bmatrix} 0 & \alpha c \\ \alpha c & 0 \end{bmatrix}$$

Because these matrices are isomorphic to the negative real numbers in all algebraic matters, we can unfold negative real numbers into this matrix form, and we can fold such matrices into negative numbers. Within an algebraic matrix form with negative space squeezing parameters, we can unfold the negative space squeezing parameters. For example, we can unfold the Euclidean complex numbers:

$$\begin{bmatrix} a & b \\ -b & a \end{bmatrix} \triangleleft \begin{bmatrix} a & 0 & b & 0 \\ 0 & a & 0 & b \\ 0 & b & a & 0 \\ b & 0 & 0 & a \end{bmatrix} \quad (12.15)$$

When we unfold an algebraic matrix form with negative space squeezing parameters, the larger matrix might not be multiplicatively closed because the block multiplication property of matrices might bring about a change of form from the 'negative number blocks'.

Digression: The hyperbolic complex numbers folded

Matrices of the appropriate form can be folded into a negative real number.

$$\begin{bmatrix} 0 & b \\ b & 0 \end{bmatrix} \triangleright -b \quad (12.16)$$

More than this, we can fold matrices of the form:

$$\begin{bmatrix} a & b \\ b & a \end{bmatrix} = \left(\begin{bmatrix} a & 0 \\ 0 & a \end{bmatrix} + \begin{bmatrix} 0 & b \\ b & 0 \end{bmatrix} \right) \triangleright a - b \quad (12.17)$$

$$a + \hat{r}b \equiv \begin{bmatrix} a & b \\ b & a \end{bmatrix} \triangleright a - b \quad (12.18)$$

This is consistent with $\hat{r} = \sqrt{+1}$ if we take $\hat{r}$ to be minus unity. However, if we reduce the $a-b$ object to a single real number, we lose a dimension - two dimensions appear to fold into one. Within the 2×2 matrix, it makes no sense to subtract one element from another.

A Chain of Algebras:

The $C_{4,2} L^1 H^3_{(j=1, k=1, l=1)}$ algebra is not the only 4-dimensional algebra. For example, by setting the space squeezing parameters as $\{j = -1, k = -1, l = 1\}$, we get:

$$C_{4.2}L^1E^3{}_{(j=-1,k=-1,l=1)}: \begin{bmatrix} a & c & b & d \\ -c & a & -d & b \\ d & -b & a & c \\ b & d & -c & a \end{bmatrix} \qquad (12.19)$$

This algebra can be folded:

$$\begin{bmatrix} a & c & b & d \\ -c & a & -d & b \\ d & -b & a & c \\ b & d & -c & a \end{bmatrix} \triangleright \begin{bmatrix} a+\hat{i}b & c+\hat{i}d \\ d-\hat{i}c & a+\hat{i}b \end{bmatrix} \triangleright a+b\hat{s}+c\hat{t}+d\hat{u} \quad (12.20)$$

It can be unfolded:

$$\begin{bmatrix} a & c & b & d \\ -c & a & -d & b \\ d & -b & a & c \\ b & d & -c & a \end{bmatrix} \triangleleft \begin{bmatrix} a & 0 & c & 0 & b & 0 & d & 0 \\ 0 & a & 0 & c & 0 & b & 0 & d \\ 0 & c & a & 0 & 0 & d & b & 0 \\ c & 0 & 0 & a & d & 0 & 0 & b \\ d & 0 & 0 & b & a & 0 & c & 0 \\ 0 & d & b & 0 & 0 & a & 0 & c \\ b & 0 & d & 0 & 0 & c & a & 0 \\ 0 & b & 0 & d & c & 0 & 0 & a \end{bmatrix} \qquad (12.21)$$

This is the $C_8 L^1 H^7{}_{(j=1,\ldots,p=1)}$ algebra (with some variables equal to zero).

Thus, we have a chain of algebras. We start with the 2-dimensional hyperbolic complex numbers, and we introduce a negative space squeezing parameter thereby forming the Euclidean complex numbers. We unfold the Euclidean complex numbers into the $C_{4.2}L^1H^3{}_{(j=1,k=1,l=1)}$ algebra. We introduce a set of negative space squeezing parameters and unfold the matrix and get the $C_8 L^1 H^7{}_{(j=1,\ldots,p=1)}$ algebra.

Of course, we cannot unfold a 2-dimensiional algebra into a 3-dimensional algebra. The algebras must double in size at each unfolding.

Section 12

Unfolding the Euclidean Complex Numbers:

In a previous chapter, we unfolded the Euclidean complex numbers and thereby removed the negative space squeezing parameter. The unfolded form of the Euclidean complex numbers is the matrix:

$$\begin{bmatrix} a & c \\ -c & a \end{bmatrix} \triangleleft \begin{bmatrix} a & 0 & c & 0 \\ 0 & a & 0 & c \\ 0 & c & a & 0 \\ c & 0 & 0 & a \end{bmatrix} \qquad (12.22)$$

Matrices of this form are not multiplicatively closed. They expand under multiplication to become the 4-dimensional $C_{4,2} L^1 H^3_{(j=1, k=1, l=1)}$ algebra.

$$\begin{bmatrix} a & 0 & c & 0 \\ 0 & a & 0 & c \\ 0 & c & a & 0 \\ c & 0 & 0 & a \end{bmatrix} \begin{bmatrix} e & 0 & g & 0 \\ 0 & e & 0 & g \\ 0 & g & e & 0 \\ g & 0 & 0 & e \end{bmatrix} = \begin{bmatrix} ae & cg & ag+ce & 0 \\ cg & ae & 0 & ag+ce \\ 0 & ag+ce & ae & cg \\ ag+ce & 0 & cg & ae \end{bmatrix}$$

$$\left(a+\hat{i}c\right)\left(e+\hat{i}g\right) = \left(ae-cg\right)+\hat{i}\left(ag+ce\right)$$

(12.23)

$$\begin{bmatrix} l & 0 & m & 0 \\ 0 & l & 0 & m \\ 0 & m & l & 0 \\ m & 0 & 0 & l \end{bmatrix} \begin{bmatrix} ae & cg & ag+ce & 0 \\ cg & ae & 0 & ag+ce \\ 0 & ag+ce & ae & cg \\ ag+ce & 0 & cg & ae \end{bmatrix} =$$

$$\begin{bmatrix} ael & cgl+agm+cem & agl+cel+aem & cgm \\ cgl+agm+cem & ael & cgm & agl+cel+aem \\ cgm & agl+cel+aem & ael & cgl+agm+cem \\ agl+cel+aem & cgm & cgl+agm+cem & ael \end{bmatrix}$$

(12.24)

Folding this gives:

186

$$\begin{bmatrix} ael & cgl+agm+cem & agl+cel+aem & cgm \\ cgl+agm+cem & ael & cgm & agl+cel+aem \\ cgm & agl+cel+aem & ael & cgl+agm+cem \\ agl+cel+aem & cgm & cgl+agm+cem & ael \end{bmatrix}$$

$$\triangleright \begin{bmatrix} ael-cgl-agm-cem & agl+cel+aem-cgm \\ -(agl+cel+aem-cgm) & ael-cgl-agm-cem \end{bmatrix}$$

(12.25)

We have:

$$(a+\hat{i}c)(e+\hat{i}g)(l+\hat{i}m)$$
$$=(ael-cgl-agm-cem)+\hat{i}(agl+cel+aem-cgm)$$
(12.26)

The $C_{4.2}\overset{1}{L}\overset{3}{H}_{(j=1,k=1,l=1)}$ algebra is, of course, closed under multiplication.

The polar form of the $C_{4.2}\overset{1}{L}\overset{3}{H}_{(j=1,k=1,l=1)}$ algebra is:

$$\exp\left(\begin{bmatrix} a & b & c & d \\ b & a & d & c \\ d & c & a & b \\ c & d & b & a \end{bmatrix}\right) =$$

$$\mathrm{PROD}\left(\begin{bmatrix} h & 0 & 0 & 0 \\ 0 & h & 0 & 0 \\ 0 & 0 & h & 0 \\ 0 & 0 & 0 & h \end{bmatrix}\begin{bmatrix} \cosh(b) & \sinh(b) & 0 & 0 \\ \sinh(b) & \cosh(b) & 0 & 0 \\ 0 & 0 & \cosh(b) & \sinh(b) \\ 0 & 0 & \sinh(b) & \cosh(b) \end{bmatrix}\right.$$

$$\begin{bmatrix} AH_4(c) & CH_4(c) & BH_4(c) & DH_4(c) \\ CH_4(c) & AH_4(c) & DH_4(c) & BH_4(c) \\ DH_4(c) & BH_4(c) & AH_4(c) & CH_4(c) \\ BH_4(c) & DH_4(c) & CH_4(c) & AH_4(c) \end{bmatrix}$$

$$\left.\begin{bmatrix} AH_4(d) & CH_4(d) & DH_4(d) & BH_4(d) \\ CH_4(d) & AH_4(d) & BH_4(d) & DH_4(d) \\ BH_4(d) & DH_4(d) & AH_4(d) & CH_4(d) \\ DH_4(d) & BH_4(d) & CH_4(d) & AH_4(d) \end{bmatrix}\right)$$

(12.27)

Section 12

We fold this into 2×2 matrices.

$$PROD \begin{pmatrix} \begin{bmatrix} e^a & 0 \\ 0 & e^a \end{bmatrix} \begin{bmatrix} \cosh(b) - \sinh(b) & 0 \\ 0 & \cosh(b) - \sinh(b) \end{bmatrix} \\ \begin{bmatrix} AH_4(c) - CH_4(c) & BH_4(c) - DH_4(c) \\ DH_4(c) - BH_4(c) & AH_4(c) - CH_4(c) \end{bmatrix} \\ \begin{bmatrix} AH_4(d) - CH_4(d) & DH_4(d) - BH_4(d) \\ BH_4(d) - DH_4(d) & AH_4(d) - CH_4(d) \end{bmatrix} \end{pmatrix}$$

$$= \begin{bmatrix} e^a & 0 \\ 0 & e^a \end{bmatrix} \begin{bmatrix} e^b & 0 \\ 0 & e^b \end{bmatrix} \begin{bmatrix} \cos(c) & \sin(c) \\ -\sin(c) & \cos(c) \end{bmatrix} \begin{bmatrix} \cos(d) & -\sin(d) \\ \sin(d) & \cos(d) \end{bmatrix}$$

$$= \begin{bmatrix} e^{a \cdot b} & 0 \\ 0 & e^{a \cdot b} \end{bmatrix} \begin{bmatrix} \cos(c-d) & \sin(c-d) \\ -\sin(c-d) & \cos(c-d) \end{bmatrix}$$

$$(12.28)$$

Two rotation matrices have combined into a single rotation matrix, and one has turned into a distance. Depolarizing it gives:

$$\begin{bmatrix} a-b & c-d \\ -(c-d) & a-b \end{bmatrix} \quad (12.29)$$

We could have obtained this directly by folding the $C_{4,2}L^1H^3_{(j=1,k=1,l=1)}$ algebra:

$$\begin{bmatrix} a & b & c & d \\ b & a & d & c \\ d & c & a & b \\ c & d & b & a \end{bmatrix} \triangleright \begin{bmatrix} a-b & c-d \\ -(c-d) & a-b \end{bmatrix} \quad (12.30)$$

The $C_{4,2}L^1H^3_{(j=1,k=1,l=1)}$ algebra contains a 2-dimensional sub-algebra. This sub-algebra is the hyperbolic complex numbers (with overblown representation):

$$\begin{bmatrix} a & b & 0 & 0 \\ b & a & 0 & 0 \\ 0 & 0 & a & b \\ 0 & 0 & b & a \end{bmatrix} \quad (12.31)$$

The Euclidean complex number algebra is a folded form of the $C_{4,2}L^1H^3_{(j=1,k=1,l=1)}$ algebra. However, the folded form of an algebra is not the same thing as the unfolded form of that algebra. The axes of the unfolded algebra are orthogonal; this is not so in the folded form. The folding has arranged the axes in a particular way – it has paired them. The sub-algebra is not apparent, and the distance functions are changed.

The Euclidean complex numbers contain the 2-dimensional Euclidean space. The $C_{4.2}L^1H^3_{(j=1,k=1,l=1)}$ algebra contains 4-dimensional space. This 4-dimensional space is manifested as 2-dimensional Euclidean space.

That the Euclidean complex numbers are a folded form of a 4-dimensional space is consistent with the fact that the unit sub-matrices of the Euclidean complex numbers form the group C_4.

Theorem:
The Euclidean complex numbers are a folded form of the 4-dimensional C_4 algebra.
Proof:
   See above.

Unfolding in General:

2nd edition note: It is still not understood whether or not folding and unfolding plays a significant role in our physical universe. It does offer an explanation for why we do not see 8-dimensional spaces, and it might be connected to the fact that a Dirac spinor has eight independent variables when it really ought to have only four.

In Section 5, we discovered that any $n \times n$ matrix that has n variables arranged so that each variable appears once and only once in every row and in every column and has the same variable as every element on the leading diagonal is an algebraic matrix form if it is multiplicatively commutative. This matrix form is guaranteed non-singular in its polar form. Clearly, any unfolded algebraic matrix form satisfies these requirements (with half of the variables at zero) if it is multiplicatively commutative and it is not a sub-algebra of a higher dimensioned algebra. (If the unfolded algebraic matrix form is a sub-algebra of the larger algebraic matrix form, it will not multiplicatively expand into the larger algebra.)

Any unfolded algebraic matrix form can be arranged as 2×2 blocks, one for each of the elements of the folded form. Because there was at least one negative space squeezing parameter, at least one of these blocks will be of the matrix form of a negative number. Such a block will multiply to form a real number matrix. The block multiplication property of matrices ensures

Section 12

that this property will be a property of the larger matrix. Thus it is that an unfolded algebraic matrix form cannot be a sub-algebra of the larger matrix.

Each 2×2 block of the unfolded algebraic matrix form is of the form of either:

$$\left\{ \begin{bmatrix} a & 0 \\ 0 & a \end{bmatrix}, \begin{bmatrix} 0 & b \\ b & 0 \end{bmatrix} \right\} \tag{12.32}$$

Both these matrix forms are multiplicatively commutative with themselves. Since the algebraic matrix forms with the negative space squeezing parameters are multiplicatively commutative, and the blocks that we have unfolded the elements of that matrix into are multiplicatively commutative, the unfolded algebraic matrix form must be multiplicatively commutative.

Thus, every time we unfold an algebraic matrix form with negative space squeezing parameters, we will get an algebraic matrix form that has no negative space squeezing parameters. There is only one such algebra (usually with many algebraically isomorphic matrix forms). Thus, every one of the algebraic matrix forms with negative space squeezing parameters of a particular size will unfold into the same algebra. In the case of the three 3-dimensional algebraic matrix forms with negative space squeezing parameters, they all unfold into the $C_6 L^1 H^5_{(j=1,\ldots,n=1)}$ algebra. All the 4-dimensional algebraic matrix forms with negative space squeezing parameters unfold into the $C_8 L^1 H^7_{(j=1,\ldots,p=1)}$ algebra.

So it is that the n-dimensional algebraic matrix forms with negative space squeezing parameters are folded forms of the $2n$-dimensional algebraic matrix forms with all positive space squeezing parameters.

When we fold a $2n$-dimensional algebraic matrix form into a n-dimensional algebraic matrix form, the elements of the n-dimensional algebraic matrix form are of the form: $(a-c)$. This mirrors the simple-trig functions where we have:

$$\cos(\theta) = AH_4(\theta) - CH_4(\theta) \tag{12.33}$$

The n-dimensional algebraic matrix forms with negative space squeezing parameters have unit sub-matrices that form the group C_{2n}. This is consistent with them being considered to be $2n$-dimensional algebras.

Thus, we effectively have one natural algebra for each cyclic group. The cross products of these groups have cross-product algebras associated with them.

Theorem:
There is one and only one natural algebra for every cyclic group.
Proof:
See above.

The natural algebra associated with a particular cyclic group has a geometric space within it. This space is always an H-type space with H-type trigonometric functions because the natural algebra has no negative space squeezing parameters.

The above theorem is a substantial simplification of the natural algebras. We now have it that a number is a cyclic group is an algebra is a geometric space is a algebraic vector product space, and that there is one, and only one, such number of order n.

A Farewell to Negative Numbers:

In non-matrix notation, the Euclidean complex numbers include the imaginary number $\hat{i}$. Clifford algebraists see this entity as a bivector; most mathematicians see it as a number of equal status to any real number. However, when we write the Euclidean complex numbers in matrix form:

$$\begin{bmatrix} a & b \\ -b & a \end{bmatrix} \quad (12.34)$$

we see that it is nothing more than a notational device for keeping track of the matrix configuration:

$$\begin{bmatrix} 0 & 1 \\ -1 & 0 \end{bmatrix} \quad (12.35)$$

Similarly, when we unfold a matrix containing negative real numbers, we see that the minus sign is nothing more than a notational device for keeping track of the matrix configuration:

$$\begin{bmatrix} 0 & 1 \\ 1 & 0 \end{bmatrix} \quad (12.36)$$

There stands the argument; let the reader decide its merit.

Theorem (contentious):

Section 12

> Negative numbers are no more than a notational device.
> **Proof:**
> See above

The above theorem does not prevent us from doing subtraction of real numbers provided the result is positive or zero. We knew this when we were at infant school where we learnt things like "three take-away five – can't do". It is only education that has led us to think that we can do such sums.

Within the physical universe, nothing exists in negative quantities[111], and so we ought to be glad to be rid of negative numbers.

Without negative numbers, a spatial axis runs from zero to plus infinity rather than from minus infinity to plus infinity. Thus it is that the '2-dimensional' Euclidean plane is 4-dimensional with axes aligned in pairs rather than orthogonally.

The axioms of an algebraic field include an axiom requiring the inclusion of additive inverses. This axiom is independent of all other axioms (that is why it is an axiom), and it can be removed without upsetting anything. If we do this, we still have an intact algebraic structure that we might refer to as a positive definite algebraic field. In a positive definite algebraic field, we still have subtraction but we view it as a form of addition – subtracting three from five expresses the same relation as adding three and two. Of course, we cannot do subtraction if the result is less than zero.

> 2nd edition note: We now know that it is very sensible to discard the axiom including additive inverses on the real axis (negative real numbers). Only the Euclidean complex numbers and the two types of quaternions satisfy this axiom. The vast majority of spinor algebras are, like the hyperbolic complex numbers which is the space-time of special relativity, without additive inverses on the real axis. We prefer reality over invented axioms.

---

[111] Electrons have a positive amount of negative electric charge.

Concluding Remarks

The existence of the natural spaces compels us to radically rethink our concepts of space.

2nd edition note: Never was a truer sentence written. We now know that the complex number spaces are spinor spaces. We know that the non-commutative algebras that derive from the $C_2 \times C_2$ groups hold electromagnetism and the 4-dimensional space-time in which we sit. We believe they also hold quantum gravity (as gravito-electromagnetism with anti-gravity) and classical general relativity, but that is not yet understood properly. These spaces contain double cover spinors, and so we see the intrinsic spin of the fermions as rotation in these spaces. We believe that the space squeezing parameters are really the physical constants of the universe. We now largely understand why we live in the universe as we see it. All of this has fallen out of our radical rethink of our concepts of space – See the two books listed in the introduction to this edition.

Appendices

Appendix 1

Axioms of a Metric:

A function of the form: $\text{distance} = f(a,b,...)$ can be used as a metric if:

i) The distance from a point to itself is zero: $d(a,a) = 0$
ii) The distance from a to b is the same as the distance from b to a: $d(a,b) = d(b,a)$
iii) The distance from a to b via c is greater than or equal to the distance from a to b (think of a triangle): $d(a,b) \le d(a,c) + d(c,b)$ with equality only if c lies on the straight line between a and b.

The second of these axioms assumes one can travel both from a to b and from b to a. Although this is possible in 2-dimensional Euclidean space, it might not be possible in a 2-dimensional space-time.

The third of these is violated by Minkowski space-time which is the space of special relativity and hence of reality. Minkowski space-time has the distance function: $x^2 - y^2 = h^2$ where h is the hypotenuse of the triangle. Clearly, the length of the hypotenuse is less than the length of the x side.

Axioms of an Inner Product:

An inner product of two vectors is denoted $\langle \underline{a}, \underline{b} \rangle$. An example is the dot product:

$$\langle \underline{a}, \underline{b} \rangle = \underline{a} \bullet \underline{b} = a_1 b_1 + a_2 b_2 + a_3 b_3 \qquad (13.1)$$

The inner product of two vectors is a projection function that produces the length of one vector projected on to the other. To qualify as an inner product, a coupling of vectors needs to satisfy:

1) $\langle \underline{a}, \underline{b} \rangle = \overline{\langle \underline{b}, \underline{a} \rangle}$ where the line over the top denotes the conjugate.
2) $\langle \lambda \underline{a}, \underline{b} \rangle = \lambda \langle \underline{a}, \underline{b} \rangle : \lambda \in \mathbb{R}$
3) $\langle \underline{a} + \underline{c}, \underline{b} \rangle = \langle \underline{a}, \underline{b} \rangle + \langle \underline{c}, \underline{b} \rangle$

4) $\langle \underline{a},\underline{a}\rangle \geq 0$ & $\langle \underline{a},\underline{a}\rangle = 0$ only when $\underline{a} = 0$

Appendix 2

The Field Axioms:

We take a set of mathematical objects. This set of mathematical objects together with operations of addition and multiplication are an algebraic field if they satisfy the following axioms.

F1: Addition is commutative
F2: Addition is associative
F3: Inclusion of the additive identity
F4: Inclusion of additive inverses
F5: Additive closure
F6: Multiplicative closure
F7: Inclusion of multiplicative inverses
F8: Inclusion of the multiplicative identity
F9: Multiplication is associative
F10: Multiplication is distributive over addition
F11: Multiplication is commutative
F12: The multiplicative identity is not the same as the additive identity
F13: The product of the additive identity with any object in the set is the additive identity
F14: There are no zero divisors

Inclusion of the additive identity means that the additive identity is one of the mathematical objects in the set; similar considerations apply when ever the word 'inclusion' appears. Additive closure means that the set of mathematical objects includes the sum of any two of the mathematical objects; similar considerations apply whenever the word 'closure' appears. Zero divisors are non-zero objects that when multiplied together produce the additive identity, zero; for example:

$$\begin{bmatrix} 4 & -4 \\ 2 & -2 \end{bmatrix}\begin{bmatrix} 1 & 3 \\ 1 & 3 \end{bmatrix} = \begin{bmatrix} 0 & 0 \\ 0 & 0 \end{bmatrix} \qquad (13.2)$$

Appendices

Within this Work:

In our case, this set of mathematical objects is always a set of matrices, and the operations of addition and multiplication are always matrix addition and matrix multiplication. Thus, in our case, we need not consider many of the above axioms because sets of matrices always satisfy them.

The determinant of the zero matrix is zero, and, since determinants are such that $\det(A)\det(B) = \det(AB)$, we can avoid zero divisors by avoiding matrices with zero determinant – singular matrices. (We also need to avoid singular matrices to include multiplicative inverses.) Thus, within this work, we need consider only:

The Axioms of an Algebraic Matrix Field:[112]

F3: Inclusion of the additive identity, $\begin{bmatrix} 0 & 0 \\ 0 & 0 \end{bmatrix}$

F4: Inclusion of additive inverses

F5: Additive Closure

F6: Multiplicative Closure

F7: Inclusion of multiplicative inverses

F8: Inclusion of the multiplicative identity, $\begin{bmatrix} 1 & 0 \\ 0 & 1 \end{bmatrix}$.

F11: Multiplication is commutative

F14: There are no non-singular matrices in the set

If F11 is satisfied, then we can always calculate the polar form of an algebra. The nature of the polar form is such that it allows no singular matrices, and so we can define the algebra to be the set of matrices that equate to the set of polar form matrices and ignore F14. The polar form automatically includes the multiplicative identity, multiplicative inverses and is multiplicatively closed; thus we can ignore F6, F7, and F8.

Appendix 3

---

[112] We demonstrate with 2×2 matrices, but the same applies to larger matrices

General Flat Algebraic Matrix Fields:

This is a list of the general form of all flat algebraic matrix forms for $n \times n$ matrices for n from 2 to 4. We also include some general forms for $n = 5, 6$. This includes only the algebraic matrix forms that have n independent variables. We ignore the restrictions necessary to avoid singular matrices; in polar form, any necessary restrictions are automatically forced on to the algebra.

The names appelled to these algebras are formed by:

i) The name of the group that underlies the algebra. If there are more than one algebra derived from a particular group (because sub-groups are generated by different elements of the group), then these different forms are denoted by a decimal point and a number added to the name of the group. Eg: $C_{4.1}$

ii) The polar form of the algebra (if known). Eg: $L^1 H^3$. The L is the (always present) length matrix. The H stands for a rotation matrix with H-type simple-trig functions within it. The E stands for a rotation matrix with E-type simple-trig functions within it.

iii) The subscripted values of the space squeezing parameters that generate the flat algebra. Eg: $_{(j=1, k=-1)}$

2 Dimensions:

The general 2-dimensional algebraic matrix form is:

$$\begin{bmatrix} a & b \\ jb & a \end{bmatrix} \qquad (13.3)$$

From which we get two flat algebras by taking $j = \{+1, -1\}$:

$$\left\{ \mathbb{C} = \begin{bmatrix} a & b \\ -b & a \end{bmatrix}, \mathbb{S} = \begin{bmatrix} a & b \\ b & a \end{bmatrix} \right\} \qquad (13.4)$$

Calculation of 3-dimensional General Form:
We start with:

$$\begin{bmatrix} a & b & c \\ jc & a & kb \\ lb & mc & a \end{bmatrix}^2 = \begin{bmatrix} a^2 + bc(j+l) & & \\ & a^2 + bc(j+km) & \\ & & a^2 + bc(l+km) \end{bmatrix}$$

$$(13.5)$$

Appendices

> We require the second matrix to be of the same form as the first matrix. This gives: $l = j, m = \dfrac{j}{k}$

3 Dimensions:

The general 3-dimensional algebraic matrix form is:

$$C_3 = \begin{bmatrix} a & b & b \\ jc & a & kb \\ jb & \dfrac{j}{k}c & a \end{bmatrix} : k \neq 0 \qquad (13.6)$$

From which we get four flat algebras by taking $\{j,k\} = \{+1,-1\}$:

$$\left\{ C_3 L^1 H^2_{(j=1,k=1)} = \begin{bmatrix} a & b & c \\ c & a & b \\ b & c & a \end{bmatrix}, C_3 L^1 E^2_{(j=1,k=-1)} = \begin{bmatrix} a & b & c \\ c & a & -b \\ b & -c & a \end{bmatrix}, \right.$$

$$\left. C_3 L^1 E^1 H^1_{(j=-1,k=1)} = \begin{bmatrix} a & b & c \\ -c & a & b \\ -b & -c & a \end{bmatrix}, C_3 L^1 E^1 H^1_{(j=-1,k=-1)} = \begin{bmatrix} a & b & c \\ -c & a & -b \\ -b & c & a \end{bmatrix} \right\}$$

(13.7)

4 Dimensions:

The 0.1 general 4-dimensional algebraic matrix form with is:

$$C_{4.1} = \begin{bmatrix} a & b & c & d \\ jd & a & kb & lc \\ \dfrac{jl}{k}c & \dfrac{j}{k}d & a & lb \\ jb & \dfrac{j}{k}c & \dfrac{j}{l}d & a \end{bmatrix} : \{k,l\} \neq 0 \qquad (13.8)$$

From which we get eight flat algebras by taking $\{j,k,l\} = \{+1,-1\}$:

(13.9)

$$\left\{ \begin{aligned}
C_{4.1}L^1H^3_{(j=1,k=1,l=1)} &= \begin{bmatrix} a & b & c & d \\ d & a & b & c \\ c & d & a & b \\ b & c & d & a \end{bmatrix}, C_{4.1}L^1E^3_{(j=1,k=1,l=-1)} = \begin{bmatrix} a & b & c & d \\ d & a & b & -c \\ -c & d & a & -b \\ b & c & -d & a \end{bmatrix}, \\
C_{4.1}L^1E^3_{(j=1,k=-1,l=1)} &= \begin{bmatrix} a & b & c & d \\ d & a & -b & c \\ -c & -d & a & b \\ b & -c & d & a \end{bmatrix}, C_{4.1}L^1H^3_{(j=1,k=-1,l=-1)} = \begin{bmatrix} a & b & c & d \\ d & a & -b & -c \\ c & -d & a & -b \\ b & -c & -d & a \end{bmatrix}, \\
C_{4.1}L^1E^3_{(j=-1,k=1,l=1)} &= \begin{bmatrix} a & b & c & d \\ -d & a & b & c \\ -c & -d & a & b \\ -b & -c & -d & a \end{bmatrix}, C_{4.1}L^1H^3_{(j=-1,k=1,l=-1)} = \begin{bmatrix} a & b & c & d \\ -d & a & b & -c \\ c & -d & a & -b \\ -b & -c & d & a \end{bmatrix}, \\
C_{4.1}L^1H^3_{(j=-1,k=-1,l=1)} &= \begin{bmatrix} a & b & c & d \\ -d & a & -b & c \\ c & d & a & b \\ -b & c & -d & a \end{bmatrix}, C_{4.1}L^1E^3_{(j=-1,k=-1,l=-1)} = \begin{bmatrix} a & b & c & d \\ -d & a & -b & -c \\ -c & d & a & -b \\ -b & c & d & a \end{bmatrix}
\end{aligned} \right.$$

(13.10)

The 0.2 general 4-dimensional algebraic matrix form is:

$$C_{4.2} = \begin{bmatrix} a & b & c & d \\ jb & a & kd & \frac{j}{k}c \\ ld & \frac{l}{k}c & a & \frac{j}{k}b \\ lc & \frac{kl}{j}d & kb & a \end{bmatrix} : \{j,k\} \neq 0 \in \mathbb{R} \qquad (13.11)$$

From which we get eight flat algebras by taking $\{j,k,l\} = \{+1,-1\}$:

$$\left\{\begin{matrix} C_{4.2}L^{1}_{(j=1,k=1,l=1)}=\begin{bmatrix} a & b & c & d \\ b & a & d & c \\ d & c & a & b \\ c & d & b & a \end{bmatrix}, C_{4.2}L^{1}_{(j=-1,k=1,l=1)}=\begin{bmatrix} a & b & c & d \\ -b & a & d & -c \\ d & c & a & -b \\ c & -d & b & a \end{bmatrix}, \\ C_{4.2}L^{1}_{(j=1,k=-1,l=1)}=\begin{bmatrix} a & b & c & d \\ b & a & -d & -c \\ d & -c & a & -b \\ c & -d & -b & a \end{bmatrix}, C_{4.2}L^{1}_{(j=1,k=1,l=-1)}=\begin{bmatrix} a & b & c & d \\ b & a & d & c \\ -d & -c & a & b \\ -c & -d & b & a \end{bmatrix}, \\ C_{4.2}L^{1}_{(j=-1,k=-1,l=1)}=\begin{bmatrix} a & b & c & d \\ -b & a & -d & c \\ d & -c & a & b \\ c & d & -b & a \end{bmatrix}, C_{4.2}L^{1}_{(j=-1,k=1,l=-1)}=\begin{bmatrix} a & b & c & d \\ -b & a & d & -c \\ -d & -c & a & -b \\ -c & d & b & a \end{bmatrix}, \\ C_{4.2}L^{1}_{(j=1,k=-1,l=-1)}=\begin{bmatrix} a & b & c & d \\ b & a & -d & -c \\ -d & c & a & -b \\ -c & d & -b & a \end{bmatrix}, C_{4.2}L^{1}_{(j=-1,k=-1,l=-1)}=\begin{bmatrix} a & b & c & d \\ -b & a & -d & c \\ -d & c & a & b \\ -c & -d & -b & a \end{bmatrix} \end{matrix}\right\} \tag{13.12}$$

The 0.3 general 4-dimensional algebraic matrix form is:

$$C_{4.3}=\begin{bmatrix} a & b & c & d \\ jc & a & kd & lb \\ jb & \frac{j}{l}d & a & \frac{j}{k}c \\ \frac{jk}{l}d & \frac{j}{l}c & kb & a \end{bmatrix} : \{k,l\} \neq 0 \in \mathbb{R} \tag{13.13}$$

From which we get eight flat algebras by taking $\{j,k,l\}=\{+1,-1\}$:

$$\left\{ C_{4,3}L^1_{(j=-1,k=-1,l=-1)} = \begin{bmatrix} a & b & c & d \\ c & a & d & b \\ b & d & a & c \\ d & c & b & a \end{bmatrix}, C_{4,3}L^1_{(j=-1,k=-1,l=-1)} = \begin{bmatrix} a & b & c & d \\ -c & a & d & b \\ -b & -d & a & -c \\ -d & -c & -b & a \end{bmatrix}, \right.$$

$$C_{4,3}L^1_{(j=-1,k=-1,l=-1)} = \begin{bmatrix} a & b & c & d \\ c & a & -d & b \\ b & d & a & -c \\ -d & c & -b & a \end{bmatrix}, C_{4,3}L^1_{(j=-1,k=-1,l=-1)} = \begin{bmatrix} a & b & c & d \\ c & a & d & -b \\ b & -d & a & -c \\ -d & -c & b & a \end{bmatrix},$$

$$C_{4,3}L^1_{(j=-1,k=-1,l=-1)} = \begin{bmatrix} a & b & c & d \\ -c & a & -d & b \\ -b & -d & a & c \\ d & -c & -b & a \end{bmatrix}, C_{4,3}L^1_{(j=-1,k=-1,l=-1)} = \begin{bmatrix} a & b & c & d \\ -c & a & d & -b \\ -b & d & a & -c \\ d & c & b & a \end{bmatrix},$$

$$\left. C_{4,3}L^1_{(j=-1,k=-1,l=-1)} = \begin{bmatrix} a & b & c & d \\ c & a & -d & -b \\ b & -d & a & -c \\ d & -c & -b & a \end{bmatrix}, C_{4,3}L^1_{(j=-1,k=-1,l=-1)} = \begin{bmatrix} a & b & c & d \\ -c & a & -d & -b \\ -b & d & a & c \\ -d & -c & -b & a \end{bmatrix} \right\}$$

(13.14)

The general 4-dimensional algebraic matrix form with basic group $C_2 \times C_2$ is:

$$C_2 \times C_2 = \begin{bmatrix} a & b & c & d \\ jb & a & \dfrac{j}{l}d & lc \\ kc & \dfrac{k}{l}d & a & lb \\ \dfrac{jk}{l^2}d & \dfrac{k}{l}c & \dfrac{j}{l}b & a \end{bmatrix} : \{l\} \neq 0 \qquad (13.15)$$

From which we get eight flat algebras by taking $\{j,k,l\} = \{+1,-1\}$:

$$\left\{\begin{array}{l}
C_2 \times C_2 L^1_{(j=1,k=1,l=1)} = \begin{bmatrix} a & b & c & d \\ b & a & d & c \\ c & d & a & b \\ d & c & b & a \end{bmatrix}, C_2 \times C_2 L^1_{(j=-1,k=1,l=1)} = \begin{bmatrix} a & b & c & d \\ -b & a & -d & c \\ c & d & a & b \\ -d & c & -b & a \end{bmatrix}, \\
C_2 \times C_2 L^1_{(j=1,k=-1,l=1)} = \begin{bmatrix} a & b & c & d \\ b & a & d & c \\ -c & -d & a & b \\ -d & -c & b & a \end{bmatrix}, C_2 \times C_2 L^1_{(j=1,k=1,l=-1)} = \begin{bmatrix} a & b & c & d \\ b & a & -d & -c \\ c & -d & a & -b \\ d & -c & -b & a \end{bmatrix}, \\
C_2 \times C_2 L^1_{(j=-1,k=-1,l=1)} = \begin{bmatrix} a & b & c & d \\ -b & a & -d & c \\ -c & -d & a & b \\ d & -c & -b & a \end{bmatrix}, C_2 \times C_2 L^1_{(j=-1,k=1,l=-1)} = \begin{bmatrix} a & b & c & d \\ -b & a & d & -c \\ c & -d & a & -b \\ -d & -c & b & a \end{bmatrix}, \\
C_2 \times C_2 L^1_{(j=1,k=-1,l=-1)} = \begin{bmatrix} a & b & c & d \\ b & a & -d & -c \\ -c & d & a & -b \\ -d & c & -b & a \end{bmatrix}, C_2 \times C_2 L^1_{(j=-1,k=-1,l=-1)} = \begin{bmatrix} a & b & c & d \\ -b & a & d & -c \\ -c & d & a & -b \\ d & c & b & a \end{bmatrix}
\end{array}\right\}$$

(13.16)

5 Dimensions:

The 0.1 general 5-dimensional algebraic matrix form is:

$$C_{5.1} = \begin{bmatrix} a & b & c & d & e \\ je & a & kb & lc & md \\ \frac{jm}{k}d & \frac{j}{k}e & a & lb & \frac{lm}{k}c \\ \frac{jm}{k}c & \frac{jm}{kl}d & \frac{j}{l}e & a & mb \\ jb & \frac{j}{k}c & \frac{j}{l}d & \frac{j}{m}e & a \end{bmatrix} : \{k,l,m \neq 0\} \qquad (13.17)$$

From which we get sixteen flat algebras by taking $\{j,k,l,m\} = \{+1,-1\}$:

Complex Numbers The Higher Dimensional Forms – 2nd Edition

6 Dimensions:

The 0.1 general 6-dimensional algebraic matrix form is:

$$C_{6.1} = \begin{bmatrix} a & b & c & d & e & f \\ jf & a & kb & lc & md & ne \\ \frac{jn}{k}e & \frac{j}{k}f & a & lb & \frac{lm}{k}c & \frac{mn}{k}d \\ \frac{jmn}{kl}d & \frac{jn}{kl}e & \frac{j}{l}f & a & mb & \frac{mn}{k}c \\ \frac{jn}{k}c & \frac{jn}{kl}d & \frac{jn}{lm}e & \frac{j}{m}f & a & nb \\ jb & \frac{j}{k}c & \frac{j}{l}d & \frac{j}{m}e & \frac{j}{n}f & a \end{bmatrix} : \{k,l,m,n \neq 0\}$$

(13.18)

From which we get thirty-two flat algebras by taking $\{j,k,l,m,n\} = \{+1,-1\}$.

The 0.2 general 6-dimensional algebraic matrix form is:

$$C_{6.2} = \begin{bmatrix} a & b & c & d & e & f \\ jb & a & kd & \frac{j}{k}c & lf & \frac{j}{l}e \\ me & \frac{lm}{j}f & a & \frac{j}{k}b & nc & \frac{kn}{l}d \\ \frac{klm}{j}f & \frac{km}{j}e & kb & a & \frac{k^2n}{j}d & \frac{kn}{l}c \\ mc & \frac{km}{j}d & \frac{m}{n}e & \frac{lm}{kn}f & a & \frac{j}{l}b \\ \frac{klm}{j}d & \frac{lm}{j}c & \frac{l^2m}{jn}f & \frac{lm}{kn}e & lb & a \end{bmatrix} : \{j,k,l,n \neq 0\}$$

(13.19)

203

Appendices

Appendix 4

Circulant Matrices:

A $n \times n$ matrix is circulant if it commutes with the appropriately sized forward shift permutation matrix. These are permutation matrices that are effectively the identity matrix shifted one place to the right. For example:

$$\left\{ \begin{bmatrix} 0 & 1 & 0 \\ 0 & 0 & 1 \\ 1 & 0 & 0 \end{bmatrix} \text{ or } \begin{bmatrix} 0 & 1 & 0 & 0 \\ 0 & 0 & 1 & 0 \\ 0 & 0 & 0 & 1 \\ 1 & 0 & 0 & 0 \end{bmatrix} \right\} \quad (13.20)$$

of which we have given the 3×3 and the 4×4 versions. The $n \times n$ forward shift matrix is a nth root of unity. Circulant matrices are 'all diagonals'; for example:

$$\left\{ \begin{bmatrix} a & b \\ b & a \end{bmatrix}, \begin{bmatrix} a & b & c \\ c & a & b \\ b & c & a \end{bmatrix}, \begin{bmatrix} a & b & c & d \\ d & a & b & c \\ c & d & a & b \\ b & c & d & a \end{bmatrix} \right\} \quad (13.21)$$

A $n \times n$ matrix is skew-circulant if it commutes with the skew forward shift matrix. The 3×3 and the 4×4 versions of skew forward shift matrices are:

$$\left\{ \begin{bmatrix} 0 & 1 & 0 \\ 0 & 0 & 1 \\ -1 & 0 & 0 \end{bmatrix} \begin{bmatrix} 0 & 1 & 0 & 0 \\ 0 & 0 & 1 & 0 \\ 0 & 0 & 0 & 1 \\ -1 & 0 & 0 & 0 \end{bmatrix} \right\} \quad (13.22)$$

The $n \times n$ skew forward shift matrix is a nth root of minus unity. Examples of skew circulant matrices are:

$$\left\{ \begin{bmatrix} a & b \\ -b & a \end{bmatrix}, \begin{bmatrix} a & b & c \\ -c & a & b \\ -b & -c & a \end{bmatrix}, \begin{bmatrix} a & b & c & d \\ -d & a & b & c \\ -c & -d & a & b \\ -b & -c & -d & a \end{bmatrix} \right\} \quad (13.23)$$

The standard form Cayley tables of all the cyclic groups are 'all diagonals'. When this table is mapped into a matrix, it becomes the circulant algebraic matrix form associated with that order cyclic group. Thus, we can write out an algebraic matrix form of a cyclic group by simply writing out an 'all diagonals' square matrix. Unfortunately, we miss the algebraically isomorphic algebraic matrix forms[113]l. We also miss the algebraic matrix forms associated with the non-cyclic abelian groups.

The determinant of the 2×2 circulant matrix is:

$$\det\left(\begin{bmatrix} a & b \\ b & a \end{bmatrix}\right) = \left(\left(\left(\sqrt[2]{+1}\right)_1\right)^2 a + \left(\sqrt[2]{+1}\right)_1 b\right)\left(\left(\left(\sqrt[2]{+1}\right)_2\right)^2 a + \left(\sqrt[2]{+1}\right)_2 b\right)$$
(13.24)

where $\left(\sqrt[2]{+1}\right)_1 = +1$ is the first square root of unity and $\left(\sqrt[2]{+1}\right)_2 = -1$ is the second square root of unity. The determinant of the 3×3 circulant matrix is:

$$\det\left(\begin{bmatrix} a & b & c \\ c & a & b \\ b & c & a \end{bmatrix}\right) = PROD\begin{pmatrix} \left(\left(\left(\sqrt[3]{+1}\right)_1\right)^3 a + \left(\left(\sqrt[3]{+1}\right)_1\right)^1 b + \left(\left(\sqrt[3]{+1}\right)_1\right)^2 c\right) \\ \left(\left(\left(\sqrt[3]{+1}\right)_2\right)^3 a + \left(\left(\sqrt[3]{+1}\right)_2\right)^1 b + \left(\left(\sqrt[3]{+1}\right)_2\right)^2 c\right) \\ \left(\left(\left(\sqrt[3]{+1}\right)_3\right)^3 a + \left(\left(\sqrt[3]{+1}\right)_3\right)^1 b + \left(\left(\sqrt[3]{+1}\right)_3\right)^2 c\right) \end{pmatrix}$$
(13.25)

where $\left(\sqrt[3]{+1}\right)_1 = +1$ is the first cube root of unity, $\left(\sqrt[3]{+1}\right)_2 = -\frac{1}{2} + \frac{\sqrt{3}}{2}\hat{i}$ is the second cube root of unity, and $\left(\sqrt[3]{+1}\right)_3 = -\frac{1}{2} - \frac{\sqrt{3}}{2}\hat{i}$ is the third cube root of unity. (Of course, these complex numbers do not exist in the 3-dimensional algebras.) The determinant of the 4×4 circulant matrix is:

$$\det\left(\begin{bmatrix} a & b & c & d \\ d & a & b & c \\ c & d & a & b \\ b & c & d & a \end{bmatrix}\right)$$
(13.26)

---

[113] The $\{C_{4,2}, C_{4,3}, C_2 \times C_2\}$ algebraic matrix forms do not commute with the forward shift permutation matrix.

Appendices

$$= PROD \begin{pmatrix} \left(\left(\left(\sqrt[4]{+1}\right)_1\right)^4 a + \left(\left(\sqrt[4]{+1}\right)_1\right)^1 b + \left(\left(\sqrt[4]{+1}\right)_1\right)^2 c + \left(\left(\sqrt[4]{+1}\right)_1\right)^3 d\right) \\ \left(\left(\left(\sqrt[4]{+1}\right)_2\right)^4 a + \left(\left(\sqrt[4]{+1}\right)_2\right)^1 b + \left(\left(\sqrt[4]{+1}\right)_2\right)^2 c + \left(\left(\sqrt[4]{+1}\right)_2\right)^3 d\right) \\ \left(\left(\left(\sqrt[4]{+1}\right)_3\right)^4 a + \left(\left(\sqrt[4]{+1}\right)_3\right)^1 b + \left(\left(\sqrt[4]{+1}\right)_3\right)^2 c + \left(\left(\sqrt[4]{+1}\right)_3\right)^3 d\right) \\ \left(\left(\left(\sqrt[4]{+1}\right)_4\right)^4 a + \left(\left(\sqrt[4]{+1}\right)_4\right)^1 b + \left(\left(\sqrt[4]{+1}\right)_4\right)^2 c + \left(\left(\sqrt[4]{+1}\right)_4\right)^3 d\right) \end{pmatrix}$$

$$= PROD \begin{pmatrix} (a+b+c+d) \\ (a-b+c-d) \\ (a+\hat{ib}-c-\hat{id}) \\ (a-\hat{ib}-c+\hat{id}) \end{pmatrix}$$

(13.27)

The pattern is clear and it applies to all circulant matrices.

Theorem:[114]
The determinant of the $n \times n$ circulant matrix is:
$$\prod_{j=1}^{j=n}\left(a(\omega_j)^n + b(\omega_j)^1 + c(\omega_j)^2 + d(\omega_j)^3 + ... + \text{variable}_n(\omega_j)^{n-1}\right)$$

The eigenvalues of the $n \times n$ circulant matrix are:
$$\lambda_j = a(\omega_j)^n + b(\omega_j)^1 + c(\omega_j)^2 + d(\omega_j)^3 + ... + \text{variable}_n(\omega_j)^{n-1}$$
(13.28)

Clearly, one basic form of the algebraic matrix forms of cyclic groups is a circulant matrix.

---
[114] See http://mathworld.wolfram.com/CirculantDeterminant.html

Appendix 5.1

Polar Forms of the $C_{4.1}$ Algebras:

The general 4-dimensional algebraic matrix form of the $C_{4.1}$ algebras is:

$$\begin{bmatrix} a & b & c & d \\ jd & a & kb & lc \\ \dfrac{jl}{k}c & \dfrac{j}{k}d & a & lb \\ jb & \dfrac{j}{k}c & \dfrac{j}{l}d & a \end{bmatrix} : \{k,l\} \neq 0 \qquad (13.29)$$

From which we get eight algebras by taking $\{j,k,l\} = \{+1,-1\}$:

$$\left\{ \begin{bmatrix} a & b & c & d \\ d & a & b & c \\ c & d & a & b \\ b & c & d & a \end{bmatrix}, \begin{bmatrix} a & b & c & d \\ d & a & b & -c \\ -c & d & a & -b \\ b & c & -d & a \end{bmatrix}, \begin{bmatrix} a & b & c & d \\ d & a & -b & c \\ -c & -d & a & b \\ b & -c & d & a \end{bmatrix}, \begin{bmatrix} a & b & c & d \\ d & a & -b & -c \\ c & -d & a & -b \\ b & -c & -d & a \end{bmatrix} \right.$$

$$\left. \begin{bmatrix} a & b & c & d \\ -d & a & b & c \\ -c & -d & a & b \\ -b & -c & -d & a \end{bmatrix}, \begin{bmatrix} a & b & c & d \\ -d & a & b & -c \\ c & -d & a & -b \\ -b & -c & d & a \end{bmatrix}, \begin{bmatrix} a & b & c & d \\ -d & a & -b & c \\ c & d & a & b \\ -b & c & -d & a \end{bmatrix}, \begin{bmatrix} a & b & c & d \\ -d & a & -b & -c \\ -c & d & a & -b \\ -b & c & d & a \end{bmatrix} \right\}$$

(13.30)

1) $\{j=1, k=1, l=1\}$

$$\exp\left(\begin{bmatrix} a & b & c & d \\ d & a & b & c \\ c & d & a & b \\ b & c & d & a \end{bmatrix}\right) = \qquad (13.31)$$

Appendices

$$PROD\left(\begin{bmatrix} h & 0 & 0 & 0 \\ 0 & h & 0 & 0 \\ 0 & 0 & h & 0 \\ 0 & 0 & 0 & h \end{bmatrix}\begin{bmatrix} AH_4(b) & BH_4(b) & CH_4(b) & DH_4(b) \\ DH_4(b) & AH_4(b) & BH_4(b) & CH_4(b) \\ CH_4(b) & DH_4(b) & AH_4(b) & BH_4(b) \\ BH_4(b) & CH_4(b) & DH_4(b) & AH_4(b) \end{bmatrix}\right.$$

$$\begin{bmatrix} \cosh(c) & 0 & \sinh(c) & 0 \\ 0 & \cosh(c) & 0 & \sinh(c) \\ \sinh(c) & 0 & \cosh(c) & 0 \\ 0 & \sinh(c) & 0 & \cosh(c) \end{bmatrix}$$

$$\left.\begin{bmatrix} AH_4(d) & DH_4(d) & CH_4(d) & BH_4(d) \\ BH_4(d) & AH_4(d) & DH_4(d) & CH_4(d) \\ CH_4(d) & BH_4(d) & AH_4(d) & DH_4(d) \\ DH_4(d) & CH_4(d) & BH_4(d) & AH_4(d) \end{bmatrix}\right)$$

(13.32)

2) $\{j=1, k=1, l=-1\}$

$$\exp\left(\begin{bmatrix} a & b & c & d \\ d & a & b & -c \\ -c & d & a & -b \\ b & c & -d & a \end{bmatrix}\right) =$$

$$PROD\left(\begin{bmatrix} h & 0 & 0 & 0 \\ 0 & h & 0 & 0 \\ 0 & 0 & h & 0 \\ 0 & 0 & 0 & h \end{bmatrix}\begin{bmatrix} AE_4(b) & BE_4(b) & CE_4(b) & -DE_4(b) \\ -DE_4(b) & AE_4(b) & BE_4(b) & -CE_4(b) \\ -CE_4(b) & -DE_4(b) & AE_4(b) & -BE_4(b) \\ BE_4(b) & CE_4(b) & DE_4(b) & AE_4(b) \end{bmatrix}\right.$$

$$\begin{bmatrix} \cos(c) & 0 & \sin(c) & 0 \\ 0 & \cos(c) & 0 & -\sin(c) \\ -\sin(c) & 0 & \cos(c) & 0 \\ 0 & \sin(c) & 0 & \cos(c) \end{bmatrix}$$

$$\left.\begin{bmatrix} AE_4(d) & -DE_4(d) & -CE_4(d) & BE_4(d) \\ BE_4(d) & AE_4(d) & -DE_4(d) & CE_4(d) \\ CE_4(d) & BE_4(d) & AE_4(d) & DE_4(d) \\ -DE_4(d) & -CE_4(d) & -BE_4(d) & AE_4(d) \end{bmatrix}\right)$$

(13.33)

3) $\{j=1, k=-1, l=1\}$

$$\exp\left(\begin{bmatrix} a & b & c & d \\ d & a & -b & c \\ -c & -d & a & b \\ b & -c & d & a \end{bmatrix}\right) =$$

$$PROD \begin{pmatrix} \begin{bmatrix} h & 0 & 0 & 0 \\ 0 & h & 0 & 0 \\ 0 & 0 & h & 0 \\ 0 & 0 & 0 & h \end{bmatrix} \begin{bmatrix} AE_4(b) & BE_4(b) & -CE_4(b) & -DE_4(b) \\ -DE_4(b) & AE_4(b) & -BE_4(b) & -CE_4(b) \\ CE_4(b) & DE_4(b) & AE_4(b) & BE_4(b) \\ BE_4(b) & CE_4(b) & -DE_4(b) & AE_4(b) \end{bmatrix} \\ \begin{bmatrix} \cos(c) & 0 & \sin(c) & 0 \\ 0 & \cos(c) & 0 & \sin(c) \\ -\sin(c) & 0 & \cos(c) & 0 \\ 0 & -\sin(c) & 0 & \cos(c) \end{bmatrix} \\ \begin{bmatrix} AE_4(d) & -DE_4(d) & CE_4(d) & BE_4(d) \\ BE_4(d) & AE_4(d) & DE_4(d) & CE_4(d) \\ -CE_4(d) & -BE_4(d) & AE_4(d) & -DE_4(d) \\ -DE_4(d) & -CE_4(d) & BE_4(d) & AE_4(d) \end{bmatrix} \end{pmatrix}$$

(13.34)

4) $\{j=1, k=-1, l=-1\}$

$$\exp\left(\begin{bmatrix} a & b & c & d \\ d & a & -b & -c \\ c & -d & a & -b \\ b & -c & -d & a \end{bmatrix}\right) = \qquad (13.35)$$

Appendices

$$PROD\left(\begin{bmatrix} h & 0 & 0 & 0 \\ 0 & h & 0 & 0 \\ 0 & 0 & h & 0 \\ 0 & 0 & 0 & h \end{bmatrix}\begin{bmatrix} AH_4(b) & BH_4(b) & -CH_4(b) & DH_4(b) \\ DH_4(b) & AH_4(b) & -BH_4(b) & CH_4(b) \\ -CH_4(b) & -DH_4(b) & AH_4(b) & -BH_4(b) \\ BH_4(b) & CH_4(b) & -DH_4(b) & AH_4(b) \end{bmatrix}\right.$$

$$\begin{bmatrix} \cosh(c) & 0 & \sinh(c) & 0 \\ 0 & \cosh(c) & 0 & -\sinh(c) \\ \sinh(c) & 0 & \cosh(c) & 0 \\ 0 & -\sinh(c) & 0 & \cosh(c) \end{bmatrix}$$

$$\left.\begin{bmatrix} AH_4(d) & DH_4(d) & -CH_4(d) & BH_4(d) \\ BH_4(d) & AH_4(d) & -DH_4(d) & CH_4(d) \\ -CH_4(d) & -BH_4(d) & AH_4(d) & -DH_4(d) \\ DH_4(d) & CH_4(d) & -BH_4(d) & AH_4(d) \end{bmatrix}\right)$$

(13.36)

5) $\{j=-1, k=1, l=1\}$

$$\exp\left(\begin{bmatrix} a & b & c & d \\ -d & a & b & c \\ -c & -d & a & b \\ -b & -c & -d & a \end{bmatrix}\right) =$$

$$PROD\left(\begin{bmatrix} h & 0 & 0 & 0 \\ 0 & h & 0 & 0 \\ 0 & 0 & h & 0 \\ 0 & 0 & 0 & h \end{bmatrix}\begin{bmatrix} AE_4(b) & BE_4(b) & CE_4(b) & DE_4(b) \\ -DE_4(b) & AE_4(b) & BE_4(b) & CE_4(b) \\ -CE_4(b) & -DE_4(b) & AE_4(b) & BE_4(b) \\ -BE_4(b) & -CE_4(b) & -DE_4(b) & AE_4(b) \end{bmatrix}\right.$$

$$\begin{bmatrix} \cos(c) & 0 & \sin(c) & 0 \\ 0 & \cos(c) & 0 & \sin(c) \\ -\sin(c) & 0 & \cos(c) & 0 \\ 0 & -\sin(c) & 0 & \cos(c) \end{bmatrix}$$

$$\left.\begin{bmatrix} AE_4(d) & DE_4(d) & -CE_4(d) & BE_4(d) \\ -BE_4(d) & AE_4(d) & DE_4(d) & -CE_4(d) \\ CE_4(d) & -BE_4(d) & AE_4(d) & DE_4(d) \\ -DE_4(d) & CE_4(d) & -BE_4(d) & AE_4(d) \end{bmatrix}\right)$$

(13.37)

6) $\{j = -1, k = 1, l = -1\}$

$$\exp\left(\begin{bmatrix} a & b & c & d \\ -d & a & b & -c \\ c & -d & a & -b \\ -b & -c & d & a \end{bmatrix}\right) =$$

$$PROD \begin{pmatrix} \begin{bmatrix} h & 0 & 0 & 0 \\ 0 & h & 0 & 0 \\ 0 & 0 & h & 0 \\ 0 & 0 & 0 & h \end{bmatrix} \begin{bmatrix} AH_4(b) & BH_4(b) & CH_4(b) & -DH_4(b) \\ DH_4(b) & AH_4(b) & BH_4(b) & -CH_4(b) \\ CH_4(b) & DH_4(b) & AH_4(b) & -BH_4(b) \\ -BH_4(b) & -CH_4(b) & -DH_4(b) & AH_4(b) \end{bmatrix} \\ \begin{bmatrix} \cosh(c) & 0 & \sinh(c) & 0 \\ 0 & \cosh(c) & 0 & -\sinh(c) \\ \sinh(c) & 0 & \cosh(c) & 0 \\ 0 & -\sinh(c) & 0 & \cosh(c) \end{bmatrix} \\ \begin{bmatrix} AH_4(d) & -DH_4(d) & CH_4(d) & BH_4(d) \\ -BH_4(d) & AH_4(d) & -DH_4(d) & -CH_4(d) \\ CH_4(d) & -BH_4(d) & AH_4(d) & DH_4(d) \\ DH_4(d) & -CH_4(d) & BH_4(d) & AH_4(d) \end{bmatrix} \end{pmatrix}$$

(13.38)

7) $\{j = -1, k = -1, l = 1\}$

$$\exp\left(\begin{bmatrix} a & b & c & d \\ -d & a & -b & c \\ c & d & a & b \\ -b & c & -d & a \end{bmatrix}\right) = \qquad (13.39)$$

Appendices

$$\left(\begin{bmatrix} h & 0 & 0 & 0 \\ 0 & h & 0 & 0 \\ 0 & 0 & h & 0 \\ 0 & 0 & 0 & h \end{bmatrix} \begin{bmatrix} AH_4(b) & BH_4(b) & -CH_4(b) & -DH_4(b) \\ DH_4(b) & AH_4(b) & -BH_4(b) & -CH_4(b) \\ -CH_4(b) & -DH_4(b) & AH_4(b) & BH_4(b) \\ -BH_4(b) & -CH_4(b) & DH_4(b) & AH_4(b) \end{bmatrix} \right.$$

$$PROD \begin{bmatrix} \cosh(c) & 0 & \sinh(c) & 0 \\ 0 & \cosh(c) & 0 & \sinh(c) \\ \sinh(c) & 0 & \cosh(c) & 0 \\ 0 & \sinh(c) & 0 & \cosh(c) \end{bmatrix}$$

$$\left. \begin{bmatrix} AH_4(d) & -DH_4(d) & -CH_4(d) & BH_4(d) \\ -BH_4(d) & AH_4(d) & DH_4(d) & -CH_4(d) \\ -CH_4(d) & BH_4(d) & AH_4(d) & -DH_4(d) \\ DH_4(d) & -CH_4(d) & -BH_4(d) & AH_4(d) \end{bmatrix} \right)$$

(13.40)

8) $\{j = -1, k = -1, l = -1\}$

$$\exp \left(\begin{bmatrix} a & b & c & d \\ -d & a & -b & -c \\ -c & d & a & -b \\ -b & c & d & a \end{bmatrix} \right) =$$

$$\left(\begin{bmatrix} h & 0 & 0 & 0 \\ 0 & h & 0 & 0 \\ 0 & 0 & h & 0 \\ 0 & 0 & 0 & h \end{bmatrix} \begin{bmatrix} AE_4(b) & BE_4(b) & -CE_4(b) & DE_4(b) \\ -DE_4(b) & AE_4(b) & -BE_4(b) & CE_4(b) \\ CE_4(b) & DE_4(b) & AE_4(b) & -BE_4(b) \\ -BE_4(b) & -CE_4(b) & DE_4(b) & AE_4(b) \end{bmatrix} \right.$$

$$PROD \begin{bmatrix} \cos(c) & 0 & \sin(c) & 0 \\ 0 & \cos(c) & 0 & -\sin(c) \\ -\sin(c) & 0 & \cos(c) & 0 \\ 0 & \sin(c) & 0 & \cos(c) \end{bmatrix}$$

$$\left. \begin{bmatrix} AE_4(d) & DE_4(d) & CE_4(d) & BE_4(d) \\ -BE_4(d) & AE_4(d) & -DE_4(d) & -CE_4(d) \\ -CE_4(d) & BE_4(d) & AE_4(d) & -DE_4(d) \\ -DE_4(d) & CE_4(d) & BE_4(d) & AE_4(d) \end{bmatrix} \right)$$

(13.41)

Appendix 5.2

Polar Forms of the $C_2 \times C_2$ Algebras:

The general 4-dimensional algebraic matrix form of the $C_2 \times C_2$ algebras is:

$$\begin{bmatrix} a & b & c & d \\ jb & a & \dfrac{j}{l}d & lc \\ kc & \dfrac{k}{l}d & a & lb \\ \dfrac{jk}{l^2}d & \dfrac{k}{l}c & \dfrac{j}{l}b & a \end{bmatrix} : \{l\} \neq 0 \quad (13.42)$$

From which we get eight flat algebras by taking permutations of $\{j,k,l\} = \{+1,-1\}$:

$$\left\{ \begin{bmatrix} a & b & c & d \\ b & a & d & c \\ c & d & a & b \\ d & c & b & a \end{bmatrix}, \begin{bmatrix} a & b & c & d \\ -b & a & -d & c \\ c & d & a & b \\ -d & c & -b & a \end{bmatrix}, \begin{bmatrix} a & b & c & d \\ b & a & d & c \\ -c & -d & a & b \\ -d & -c & b & a \end{bmatrix}, \begin{bmatrix} a & b & c & d \\ b & a & -d & -c \\ c & -d & a & -b \\ d & -c & -b & a \end{bmatrix}, \right.$$

$$\left. \begin{bmatrix} a & b & c & d \\ -b & a & -d & c \\ -c & -d & a & b \\ d & -c & -b & a \end{bmatrix}, \begin{bmatrix} a & b & c & d \\ -b & a & d & -c \\ c & -d & a & -b \\ -d & c & b & a \end{bmatrix}, \begin{bmatrix} a & b & c & d \\ b & a & -d & -c \\ -c & d & a & -b \\ -d & c & -b & a \end{bmatrix}, \begin{bmatrix} a & b & c & d \\ -b & a & d & -c \\ -c & d & a & -b \\ d & c & b & a \end{bmatrix} \right\}$$

(13.43)

1) $\{j=1, k=1, l=1\}$

$$\exp\left(\begin{bmatrix} a & b & c & d \\ b & a & d & c \\ c & d & a & b \\ d & c & b & a \end{bmatrix} \right) = \quad (13.44)$$

213

Appendices

$$PROD\left(\begin{bmatrix} h & 0 & 0 & 0 \\ 0 & h & 0 & 0 \\ 0 & 0 & h & 0 \\ 0 & 0 & 0 & h \end{bmatrix}\begin{bmatrix} \cosh(b) & \sinh(b) & 0 & 0 \\ \sinh(b) & \cosh(b) & 0 & 0 \\ 0 & 0 & \cosh(b) & \sinh(b) \\ 0 & 0 & \sinh(b) & \cosh(b) \end{bmatrix}\right.$$

$$\begin{bmatrix} \cosh(c) & 0 & \sinh(c) & 0 \\ 0 & \cosh(c) & 0 & \sinh(c) \\ \sinh(c) & 0 & \cosh(c) & 0 \\ 0 & \sinh(c) & 0 & \cosh(c) \end{bmatrix}$$

$$\left.\begin{bmatrix} \cosh(d) & 0 & 0 & \sinh(d) \\ 0 & \cosh(d) & \sinh(d) & 0 \\ 0 & \sinh(d) & \cosh(d) & 0 \\ \sinh(d) & 0 & 0 & \cosh(d) \end{bmatrix}\right)$$

(13.45)

2) $\{j=-1, k=1, l=1\}$

$$\exp\left(\begin{bmatrix} a & b & c & d \\ -b & a & -d & c \\ c & d & a & b \\ -d & c & -b & a \end{bmatrix}\right) =$$

$$PROD\left(\begin{bmatrix} h & 0 & 0 & 0 \\ 0 & h & 0 & 0 \\ 0 & 0 & h & 0 \\ 0 & 0 & 0 & h \end{bmatrix}\begin{bmatrix} \cos(b) & \sin(b) & 0 & 0 \\ -\sin(b) & \cos(b) & 0 & 0 \\ 0 & 0 & \cos(b) & \sin(b) \\ 0 & 0 & -\sin(b) & \cos(b) \end{bmatrix}\right.$$

$$\begin{bmatrix} \cosh(c) & 0 & \sinh(c) & 0 \\ 0 & \cosh(c) & 0 & \sinh(c) \\ \sinh(c) & 0 & \cosh(c) & 0 \\ 0 & \sinh(c) & 0 & \cosh(c) \end{bmatrix}$$

$$\left.\begin{bmatrix} \cos(d) & 0 & 0 & \sin(d) \\ 0 & \cos(d) & -\sin(d) & 0 \\ 0 & \sin(d) & \cos(d) & 0 \\ -\sin(d) & 0 & 0 & \cos(d) \end{bmatrix}\right)$$

(13.46)

3) $\{j=1, k=-1, l=1\}$

$$\exp\left(\begin{bmatrix} a & b & c & d \\ b & a & d & c \\ -c & -d & a & b \\ -d & -c & b & a \end{bmatrix}\right) =$$

$$\text{PROD}\left(\begin{bmatrix} h & 0 & 0 & 0 \\ 0 & h & 0 & 0 \\ 0 & 0 & h & 0 \\ 0 & 0 & 0 & h \end{bmatrix}\begin{bmatrix} \cosh(b) & \sinh(b) & 0 & 0 \\ \sinh(b) & \cosh(b) & 0 & 0 \\ 0 & 0 & \cosh(b) & \sinh(b) \\ 0 & 0 & \sinh(b) & \cosh(b) \end{bmatrix}\right.$$

$$\begin{bmatrix} \cos(c) & 0 & \sin(c) & 0 \\ 0 & \cos(c) & 0 & \sin(c) \\ -\sin(c) & 0 & \cos(c) & 0 \\ 0 & -\sin(c) & 0 & \cos(c) \end{bmatrix}$$

$$\left.\begin{bmatrix} \cos(d) & 0 & 0 & \sin(d) \\ 0 & \cos(d) & \sin(d) & 0 \\ 0 & -\sin(d) & \cos(d) & 0 \\ -\sin(d) & 0 & 0 & \cos(d) \end{bmatrix}\right)$$

(13.47)

4) $\{j=1, k=1, l=-1\}$

$$\exp\left(\begin{bmatrix} a & b & c & d \\ b & a & -d & -c \\ c & -d & a & -b \\ d & -c & -b & a \end{bmatrix}\right) = \qquad (13.48)$$

Appendices

$$\left(\begin{bmatrix} h & 0 & 0 & 0 \\ 0 & h & 0 & 0 \\ 0 & 0 & h & 0 \\ 0 & 0 & 0 & h \end{bmatrix}\begin{bmatrix} \cosh(b) & \sinh(b) & 0 & 0 \\ \sinh(b) & \cosh(b) & 0 & 0 \\ 0 & 0 & \cosh(b) & -\sinh(b) \\ 0 & 0 & -\sinh(b) & \cosh(b) \end{bmatrix}\right.$$
$$PROD \begin{bmatrix} \cosh(c) & 0 & \sinh(c) & 0 \\ 0 & \cosh(c) & 0 & -\sinh(c) \\ \sinh(c) & 0 & \cosh(c) & 0 \\ 0 & -\sinh(c) & 0 & \cosh(c) \end{bmatrix}$$
$$\left.\begin{bmatrix} \cosh(d) & 0 & 0 & \sinh(d) \\ 0 & \cosh(d) & -\sinh(d) & 0 \\ 0 & -\sinh(d) & \cosh(d) & 0 \\ \sinh(d) & 0 & 0 & \cosh(d) \end{bmatrix}\right)$$
(13.49)

5) $\{j=-1, k=-1, l=1\}$

$$\exp\left(\begin{bmatrix} a & b & c & d \\ -b & a & -d & c \\ -c & -d & a & b \\ d & -c & -b & a \end{bmatrix}\right) =$$

$$\left(\begin{bmatrix} h & 0 & 0 & 0 \\ 0 & h & 0 & 0 \\ 0 & 0 & h & 0 \\ 0 & 0 & 0 & h \end{bmatrix}\begin{bmatrix} \cos(b) & \sin(b) & 0 & 0 \\ -\sin(b) & \cos(b) & 0 & 0 \\ 0 & 0 & \cos(b) & \sin(b) \\ 0 & 0 & -\sin(b) & \cos(b) \end{bmatrix}\right.$$
$$PROD \begin{bmatrix} \cos(c) & 0 & \sin(c) & 0 \\ 0 & \cos(c) & 0 & \sin(c) \\ -\sin(c) & 0 & \cos(c) & 0 \\ 0 & -\sin(c) & 0 & \cos(c) \end{bmatrix}$$
$$\left.\begin{bmatrix} \cosh(d) & 0 & 0 & \sinh(d) \\ 0 & \cosh(d) & -\sinh(d) & 0 \\ 0 & -\sinh(d) & \cosh(d) & 0 \\ \sinh(d) & 0 & 0 & \cosh(d) \end{bmatrix}\right)$$
(13.50)

6) $\{j=-1, k=1, l=-1\}$

$$\exp\left(\begin{bmatrix} a & b & c & d \\ -b & a & d & -c \\ c & -d & a & -b \\ -d & -c & b & a \end{bmatrix}\right) =$$

$$PROD \begin{pmatrix} \begin{bmatrix} h & 0 & 0 & 0 \\ 0 & h & 0 & 0 \\ 0 & 0 & h & 0 \\ 0 & 0 & 0 & h \end{bmatrix} \begin{bmatrix} \cos(b) & \sin(b) & 0 & 0 \\ -\sin(b) & \cos(b) & 0 & 0 \\ 0 & 0 & \cos(b) & -\sin(b) \\ 0 & 0 & \sin(b) & \cos(b) \end{bmatrix} \\ \begin{bmatrix} \cosh(c) & 0 & \sinh(c) & 0 \\ 0 & \cosh(c) & 0 & -\sinh(c) \\ \sinh(c) & 0 & \cosh(c) & 0 \\ 0 & -\sinh(c) & 0 & \cosh(c) \end{bmatrix} \\ \begin{bmatrix} \cos(d) & 0 & 0 & \sin(d) \\ 0 & \cos(d) & \sin(d) & 0 \\ 0 & -\sin(d) & \cos(d) & 0 \\ -\sin(d) & 0 & 0 & \cos(d) \end{bmatrix} \end{pmatrix}$$

(13.51)

7) $\{j=1, k=-1, l=-1\}$

$$\exp\left(\begin{bmatrix} a & b & c & d \\ b & a & -d & -c \\ -c & d & a & -b \\ -d & c & -b & a \end{bmatrix}\right) = \qquad (13.52)$$

217

Appendices

$$PROD\left(\begin{bmatrix} h & 0 & 0 & 0 \\ 0 & h & 0 & 0 \\ 0 & 0 & h & 0 \\ 0 & 0 & 0 & h \end{bmatrix} \begin{bmatrix} \cosh(b) & \sinh(b) & 0 & 0 \\ \sinh(b) & \cosh(b) & 0 & 0 \\ 0 & 0 & \cosh(b) & -\sinh(b) \\ 0 & 0 & -\sinh(b) & \cosh(b) \end{bmatrix} \right.$$

$$\begin{bmatrix} \cos(c) & 0 & \sin(c) & 0 \\ 0 & \cos(c) & 0 & -\sin(c) \\ -\sin(c) & 0 & \cos(c) & 0 \\ 0 & \sin(c) & 0 & \cos(c) \end{bmatrix}$$

$$\left.\begin{bmatrix} \cos(d) & 0 & 0 & \sin(d) \\ 0 & \cos(d) & -\sin(d) & 0 \\ 0 & \sin(d) & \cos(d) & 0 \\ -\sin(d) & 0 & 0 & \cos(d) \end{bmatrix}\right)$$

(13.53)

8) $\{j=-1, k=-1, l=-1\}$

$$\exp\left(\begin{bmatrix} a & b & c & d \\ -b & a & d & -c \\ -c & d & a & -b \\ d & c & b & a \end{bmatrix}\right) =$$

$$PROD\left(\begin{bmatrix} h & 0 & 0 & 0 \\ 0 & h & 0 & 0 \\ 0 & 0 & h & 0 \\ 0 & 0 & 0 & h \end{bmatrix} \begin{bmatrix} \cos(b) & \sin(b) & 0 & 0 \\ -\sin(b) & \cos(b) & 0 & 0 \\ 0 & 0 & \cos(b) & -\sin(b) \\ 0 & 0 & \sin(b) & \cos(b) \end{bmatrix} \right.$$

$$\begin{bmatrix} \cos(c) & 0 & \sin(c) & 0 \\ 0 & \cos(c) & 0 & -\sin(c) \\ -\sin(c) & 0 & \cos(c) & 0 \\ 0 & \sin(c) & 0 & \cos(c) \end{bmatrix}$$

$$\left.\begin{bmatrix} \cosh(d) & 0 & 0 & \sinh(d) \\ 0 & \cosh(d) & \sinh(d) & 0 \\ 0 & \sinh(d) & \cosh(d) & 0 \\ \sinh(d) & 0 & 0 & \cosh(d) \end{bmatrix}\right)$$

(13.54)

Glossary & Nomenclature

Algebra: A linear space together with a multiplication operation that satisfies the requisite axioms.

Algebraic Closure: An algebra is algebraically closed if every polynomial within it has roots that are also within the algebra. The Euclidean complex numbers are algebraically closed.

Algebraic Field: An algebra that satisfies the field axioms. A multiplicatively commutative division algebra

Algebraic Matrix Form: A matrix form that together with matrix multiplication satisfies the requirements to be an algebra.

Algebraic Operations: The operations of addition, multiplication, and multiplication by a scalar.

Angle Matrix: A $n \times n$ matrix that a member of the SO_n group of matrices and that has simple-trigonometric functions or nu-functions (compound trig functions), or zero, for its elements. An angle matrix is the same as a rotation matrix.

Angle Product: A way of multiplying two matrices together that is equivalent to both the inner product and the exterior product(s).

Cayley Table Standard Form: The Cayley table of a group written with the top row in alphabetical order and the identity elements in the leading diagonal.

Compound Polar Form: The form of a *n*-dimensional algebra written as a product of a length matrix and a single angle matrix. The angle matrix in the compound polar form is the product of $(n-1)$ angle matrices which have simple-trig functions for their elements. .

Glossary & Nomenclature

Compound Rotation Matrix: The single angle matrix in the compound polar form.

Compound-trig Function: An element of a compound rotation matrix. They are projections from the unit sphere of the space directly on to an axis.

Diagonal Matrix: A square matrix with all elements zero except the elements on the leading diagonal.

Dihedralions: Division algebras whose under-lying group is a dihedral one.

Distance Function: A function whose input is the co-ordinates of two points in space and whose output is the distance between those two points. It is invariant under transformation of axes and rotation.

Division Algebra: An algebra in which one can always do division – there are no zero-divisors.

E-Type Algebraic Matrix Form: An algebraic matrix form that gives rise to a simple polar form in which every angle matrix is a E-type angle matrix.

E-type angle matrix: An angle matrix that has elements which are E-type simple-trig functions.

E-type simple-trig function: A n dimensional simple-trig function. The series expansion has minus signs in it.

Exterior Product: The exterior product of two vectors. It is related to the cross product of two vectors. It is denoted with a wedge.

Flat Algebra: A natural algebra that has the space squeezing parameters $(j,k,l,....) = \pm 1$

General Polar Form: The polar form of a general algebraic matrix form

H-Type Algebraic Matrix Form:	An algebraic matrix form that gives rise to a simple polar form in which every angle matrix is a H-type angle matrix. This algebraic matrix form has no negative elements (minus signs) in it.
H-type angle matrix:	An angle matrix that has elements which are H-type simple-trig functions.
H-type simple-trig function:	A n dimensional simple-trig function. The series expansion has no minus signs in it.
Hyper-geometric functions:	Messed about versions of the exponential function. A sub-set of the hyper-geometric functions is the simple-trigonometric functions.
Simple-trig Function:	The functions that occur in the simple angle matrices of the polar form. Combined together by matrix multiplication, they form the v-functions.
Hyperbolic Complex Numbers:	One of the two flat 2-dimensional natural algebraic fields. The other is the Euclidean complex numbers.
Inner Product:	A projection function from one line in a space on to another line in that space. It is often called the dot-product or scalar-product. It is invariant under transformation of axes. See the inner product axioms in the appendices.
Leading Diagonal:	The diagonal of a square matrix or square table that runs from the top left-hand corner to the bottom right-hand corner.
Length Matrix:	A matrix that is isomorphic to a real number and that occurs in the polar form (simple or compound) of an algebra.
Linear Space:	A set of mathematical objects together with an addition operation are a linear space if they satisfy the requisite axioms. A closed set of linear transformations.

Glossary & Nomenclature

Matrix Form: A matrix in which each element is specified either as independent of all other elements or related to them by a specified function. For example:
$$\begin{bmatrix} a_{11} = \mathbb{R} & a_{12} = \mathbb{R} \\ a_{21} = -5a_{12} & a_{22} = a_{11} + 2a_{21} \end{bmatrix}$$

Metric: A distance function that satisfies the metric axioms.

n-dimensional algebra: An algebra with *n* independent variables.

Natural Algebra: An algebra that has matrix multiplication as the multiplication operation.

Natural Algebraic Field: An algebraic field that has matrix multiplication as the multiplication operation.

nu-function or v-function: A function in the compound polar form angle matrix. Also known as compound trigonometric functions. They are projection functions from the unit sphere on to the axes of the space.

Odd Parity Matrices: Matrices whose product is an element of a natural algebraic field.

Overblown Representation: The matrix representation of an algebra by a larger matrix form than is necessary to contain the algebraic operations.

Permutation Matrix: A matrix that is produced by swapping columns of the identity matrix. Each represents a permutation.

Pure Permutation Matrix: A permutation matrix with an all zeros leading diagonal.

Polar Form of an Algebra: The result of exponentiating the algebra.
Eg: $\exp(a + \hat{i}b) = r(\cos\theta + \hat{i}\sin\theta)$

Quaternions:	The quaternion division algebra of William Hamilton. It contains the Euclidean complex numbers as a sub-algebra.
Real Number Matrix:	A matrix that has all elements equal to zero except the elements on the leading diagonal and in which the leading diagonal elements are equal. These matrices are algebraically isomorphic to the real numbers.
Rotation Matrix:	*See: Angle Matrix*
Scalars:	Real numbers or matrices that are algebraically isomorphic to the real numbers.
Shadow Algebra:	A lesser dimensional representation of an algebra
Simple Polar Form:	The form of a n-dimensional algebra written as a product of a length matrix and $(n-1)$ angle matrices. Each of the angle matrices in the simple polar form has simple trigonometric functions for its elements.
Simple Rotation Matrix:	A rotation matrix in the simple polar form of an algebra.
Simple-Trig Function:	A n-ways splitting of the exponential series.
Space Squeezing Parameter:	Parameters that appear in algebraic matrix forms which have the effect of squeezing the distance function 'out of shape'.
Sub-Algebra:	An algebra that is contained within a larger algebra. For example, the real number algebra is contained within the complex number algebras.
Study numbers:	Hyperbolic complex numbers
Trig Function:	Geometrically, a projection on to an axis.

Unit Sub-Matrix:	A matrix whose elements are either zero or 1 and which occurs as a sub-matrix of an algebraic matrix form – the b matrix with $b=1$ for example. The multiplicative identity matrix is a unit sub-matrix of each algebraic matrix form.
Vector Matrix:	Another name for an algebraic matrix form. The term is used when considering the matrices as vectors.
Vector Space:	Another name for a linear space
Wedge Product:	The exterior product of two vectors.
Zero divisors:	A mathematical object, A or B, within an algebra is a zero divisor if $AB = 0$ when $A \neq 0$ & $B \neq 0$

Notation

$A_{[\text{Row, Column}]}$:   Denotes a particular element of a matrix, A.

AH_n: The first H-type simple-trig function in n dimensions. This function has a power series expansion beginning with x^0 and containing no negative terms (minus signs).

AE_n: The first E-type simple-trig function in n dimensions. This function has a power series expansion beginning with x^0 and containing an equal number of negative terms and positive terms (minus signs & plus signs).

$BH_n...ZH_n$:   The other H-type simple-trig functions in n dimensions. These functions have a power series expansion beginning with $x^{B...Z}$ and containing no negative terms (minus signs).

$BE_n...ZE_n$:   The other E-type simple-trig functions in n dimensions. These functions have a power series expansion beginning with $x^{B...Z}$ and containing an equal number of negative terms and positive terms (minus signs & plus signs).

C_n: The cyclic group of order n.

$C_n \times C_m$:   The group that is the direct product of the groups named

$C_{n.m}$:   The name appelled to the general form of an algebra.

$C_{n.m} L^1 H^p E^{(n-p-1)}_{(j=+1, k=+1,...)}$:   The name appelled to a particular algebra. If the polar form is not known, the polar form part of the notation will be simply the length matrix.

$\mathbb{C}$:   The complex numbers

det: The determinant of a matrix.

Notation

$\exp([A])$: The exponential of a matrix

$\begin{bmatrix} e & 0 \\ 0 & e \end{bmatrix}^{[A]}$: An alternative notation for the exponential of a 2 by 2 matrix

$\hat{i}$: The square root of minus unity - $\sqrt{-1}$. In general, such objects will carry a hat symbol above them.

$L^l H^n E^m$ The polar form of an algebra

$\log m_{[b]}([A])$: The logarithm of a matrix to base $[B]$ where B is a matrix that is algebraically isomorphic to the real number b and of the same size as the matrix A.

$v_{[\ ]}A(b,...z)$: The nu-functions in the compound polar form angle matrix of the algebra that appears in the subscript. They correspond to the cosine or the cosh function in 2-dimensions. They are sums of products of simple-trig functions formed by the matrix multiplication of the $(m+n)$ angle matrices in the simple polar form.

$v_{[\ ]}B...Z(b,...z)$: The nu-functions in the compound polar form angle matrix of the algebra that appears in the subscript. They are sums of products of simple-trig functions formed by the matrix multiplication of the $(m+n)$ angle matrices in the simple polar form.

OP_C The odd parity matrices associated with the Euclidean complex numbers.

OP_S The odd parity matrices associated with the hyperbolic complex numbers.

$PROD(\ )$: The product of the objects within the bracket.

p, q:	The cube roots of plus or minus unity in the non-matrix notation of the 3-dimension algebras - $a + b\hat{p} + c\hat{q}$. This is analogous to the non-matrix representation of the complex numbers.
$\mathbb{Q}$:	The quaternions of William Hamilton
$\mathbb{R}$:	The real numbers
$\mathbb{S}$:	The hyperbolic complex numbers (Study numbers)).
$\{\triangleright, \triangleleft\}$	Used to indicate equivalence under reduction from the overblown representation or from an algebraically isomorphic representation. These symbols are also used to indicate the folding and unfolding operations.
$[A] \odot [B]$	The angle product of the matrices $\{A, B\}$. The angle product contains the vector inner product and the exterior vector product(s).
$\begin{bmatrix} a \\ b \end{bmatrix} \wedge \begin{bmatrix} c \\ d \end{bmatrix}$	The exterior product of two vectors. Also known as the wedge product.

Other Books by the Same Author

The Naked Spinor – a Rewrite of Clifford Algebra

Spinors exist in Clifford algebras. In this book, we explore the nature of spinors. This book is an excellent introduction to Clifford algebra.

Complex Numbers The Higher Dimensional Forms – Spinor Algebra

In this book, we explore the higher dimensional forms of complex numbers. These higher dimensional forms are connected very closely to spinors.

Upon General Relativity

In this book, we see how 4-dimensional space-time, gravity, and electromagnetism emerge from the spinor algebras. This is an excellent and easy-paced introduction to general relativity.

From Where Comes the Universe

This is a guide for the lay-person to the physics of empty space.

Empty Space is Amazing Stuff – The Special Theory of Relativity

This book deduces the theory of special relativity from the finite groups. It gives a unique insight into the nature of the 2-dimensional space-time of special relativity.

The Nuts and Bolts of Quantum Mechanics

This is a gentle introduction to quantum mechanics for undergraduates.

Quaternions

This book pulls together the often separate properties of the quaternions. Non-commutative differentiation is covered as is non-commutative rotation and non-commutative inner products along with the quaternion trigonometric functions.

The Uniqueness of our Space-time

This book reports the finding that the only two geometric spaces within the finite groups are the two spaces that together form our universe. This is a startling finding. The nature of geometric space is explained alongside the nature of division algebra space, spinor space. This book is a catalogue of the higher dimensional complex numbers up to dimension fifteen.

Lie Groups and Lie Algebras

This book presents Lie theory from a diametrically different perspective to the usual presentation. This makes the subject much more intuitively obvious and easier to learn. Included is perhaps the clearest and simplest presentation of the true nature of the Lie group $SU(2)$ ever presented.

The Physics of Empty Space

This book presents a comprehensive understanding of empty space. The presence of 2-dimensional rotations in our 4-dimensional space-time is explained. Also included is a very gentle introduction to non-commutative differentiation. Classical electromagetism is deduced from the quaternions.

The Electron

This book presents the quantum field theory view of the electron and the neutrino. This view is radically different from the classical view of the electron presented in most schools and colleges. This book gives a very clear exposition of the Dirac equation including the quaternion rewrite of the Dirac equation. This is an excellent introduction to particle physics for students prior to university, during university and after university courses in physics.

The Quaternion Dirac Equation

This small book (only 40 pages) presents the quaternion form of the Dirac equation. The neutrino mass problem is solved and we gain an explanation of why neutrinos are left-chiral. Much of the material in this book is drawn from 'The Electron'; this material is presented concisely and inexpensively for students already familiar with QFT.

An Essay on the Nature of Space-time

This small and inexpensive volume presents a view of the nature of empty space without the detailed mathematics. The expanding universe and dark energy is discussed.

Elementary Calculus from an Advanced Standpoint

This book rewrite the calculus of the complex numbers in a way that covers all division algebras and makes all continuous complex functions differentiable and integrable. Non-commutative differentiation is covered. Gauge covariant differentiation is covered as is the covariant derivative of general relativity.

Other Books by the Same Author

Even Mathematicians and Physicists make Mistakes

This book points out what seems to be several important errors of modern physics and modern mathematics. Errors like the misunderstanding of rotation, the failure to teach the higher dimensional complex numbers in most universities, and the mathematical inconsistency of the Dirac equation and some casual errors are discussed. These errors are set in their historical circumstances and there is discussion about why they happened and the consequences of their happening. There is also an interesting chapter on the nature of mathematical proof within our society, and several famous proofs are discussed (without the details).

Finite Groups – A Simple Introduction

This book introduces the reader to finite group theory. Many introductory books on finite groups bury the reader in geometrical examples or in other types of groups and lose the central nature of a finite group. This book sticks firmly with the permutation nature of finite groups and elucidates that nature by the extensive use of permutation matrices. Permutation matrices simplify the subject considerably. This book is probably unique in its use of permutation matrices and therefore unique in its simplicity.

Index

1
1-dimensional space, 57

3
3-dimensional algebras, 97
3-dimensional trig funcs, 97

A
adjoint matrix, 17
algebraic closure, 55
algebraic field, 6
algebraic field extension, 13
algebraic inclusion, 13
algebraic isomorphis, 9
algebraic matrix form, 16, 65
Algebraic Motors, 38
angle product, 29
Anormal complex numbers, 38
Argand, 3
Argand diagram, hyperbolic, 39

B
bi-vector, 30
block algebras, 73
block multiplication, 8

C
Cauchy-Riemann equations, 23, 41, 114, 151
Cayley table, 76, 168
circulant matrices, 71, 200
Clifford algebra, 31, 47, 48, 53
Clifford product, 31
Clifford, William, K, 4
Cockle, 37
complex number, euclidean, 7
compound trig func, 83
compound trigonometric function, 90
conjugate, 17
cross product, 30
cross product, complex number, 27
curl, hyperbolic, 56

D
determinant, 25
differential, 128
differential, matrix, 35
differentiation cycle, 133
differentiation, matrix, 19
dihedral algebras, 54
di-hedralions, 49
distance function, 24, 26
distance function, definition, 94
divergence, 56
divergence, hyperbolic, 56
dot product, complex number, 27
Double numbers, 38
Dual numbers, 38

E
electromagnetism, 188
Euler relations, 43
exponential of a matrix, 15
exterior product, 29

F
field axioms, 61
flat algebras, 68
folding operation, 174

G
Gauss, 3
general relativity, 188
geometric space, 82
Gibbs, 4
gradient, 35
gradient, hyperbolic, 56
graphs, 129, 155
Grassman, 4
gravito-electromagnetism, 188

Index

H

Hamilton, 3
Hilbert space, 14
Hurwitz theorem, 5
hyperbolic complex number, 37, 38
hypergeometrics, 86

I

illegitimate spaces, 77
inner product, 14

L

Laplacian, 36
linear space, 6
logarithms, 32, 43, 123
Lorentz numbers, 38

M

Minkowski space-time, 43, 56
monic minimum polynomial, 13

N

names of algebras, 71
natural algebras, 7, 15
natural space, definition, 95
non-commutative algebras, 1
non-matrix notation, 115, 146
norm, 14
nu-functions, 106

O

odd parity matrices, 45
overblown representations, 12

P

parameter, space squeezing, 65
Peano, 4
permutation matrices, 163, 171
Perplex numbers, 38
polar form, hyperbolic, 41

Q

quantum gravity, 188
quaternions, 45, 49, 52

R

real-number matrices, 10
rotation, 1-dimensional, 57
rotation, definition, 91

S

scalar field, complex numbers, 35
scalar field, hyperbolic, 56
scalar multiplication, 11
Servois, 3
shadow algebra, 120, 149
simple trigonometric function, 83, 86, 128
special relativity, 43
spinor space, 1, 58
Split complex numbers, 38
Study numbers, 37
Study, Eduard, 37
super-imposition, 58

T

Tessarines, 38
trigonometric function, 25, 82, 126

V

vector, 27
Vector algebra, 4
vector calculus, 35
vector calculus, hyperbolic, 56
vector field, complex number, 35
vector field, hyperbolic, 56
vector, 3-dimensional, 124
vector, complex number, 27
vectors, hyperbolic space, 55

W

Wallis, 3
wedge product, 29
Wessel, 3

Printed in Great Britain
by Amazon